The Back Road to Crazy

The Back Road to Crazy

Stories from the Field

Edited by Jennifer Bové

THE UNIVERSITY
OF UTAH PRESS

Salt Lake City

10 09 08 07 06 05   5 4 3 2 1

 The Defiance House Man colophon is a registered trademark of the University of Utah Press. It is based upon a four-foot-tall, Ancient Puebloan pictograph (late PIII) near Glen Canyon, Utah.

"Bit" by Mark W. Moffett was originally published in a slightly different form in *Outside* (April 2002).

"Pileated Woodpeckers" by Patrick Loafman was originally published in a slightly different form in the Web journal *C/Oasis* (May 2002).

"Up Trapper Creek" by Jennifer Bové was originally published in a slightly different form as "Another Day in the Field" in *Women in Natural Resources* (Spring 2002).

"Love in the Margin: Finding Refuge in the Blue Mountains" by Joseph L. Ebersole was originally published in a slightly different form in *OE Journal* (Winter 1999).

"Tigerland" by Eric Dinerstein will be included in a slightly different form in the forthcoming book *Tigerland, and Other Unintended Destinations* by Eric Dinerstein (Washington, D.C.: Shearwater Press, 2005).

"Rollie's Catch" by Chris Smith was originally published in a slightly different form in *Gray's Sporting Journal*.

"Polar Summer" by Cameron Walker was originally published in a slightly different form in *Passionfruit: A Women's Travel Journal* (Winter 2000).

"Snapper's Kiss" by Jeff Beane was originally published in a slightly different form as part of "Sultans of Snap" in *Wildlife in North Carolina* (April 2003).

"Dependence Day" by Jeff Beane was originally published in a slightly different form in *NC Herps*, the newsletter of the North Carolina Herpetological Society (October 1996).

"The Old Camp Coffee-Pot" by Badger Clark is reprinted from *Sun and Saddle Leather*, 6th ed. (Boston: Gorham Press, 1922). It was suggested for *The Back Road to Crazy* by Steve Hayes, recently retired senior environmental scientist for the California Department of Water Resources. According to Steve, the poem is one of his favorites because "that which is bitterest to endure is always sweetest to remember."

Cover images: © Indexstock photo by John Luke; © Brand X Pictures photo by Burke/Triolo Productions; © Getty Images, James Gritz; © Getty Images, Russell Illig; © Thinkstock

LIBRARY OF CONGRESS CATALOGING-IN-PUBLICATION DATA

The back road to crazy : stories from the field / collected and edited by Jennifer Bové.
    p. cm.
  ISBN 0-87480-816-2 (pbk. : alk. paper)
  1. Biology—Field work—Anecdotes. I. Bové, Jennifer, 1973-
QH318.5.B27 2005
570'.72'3—dc22   2004020807

Contents

Introduction   1

1 Bit, *Mark W. Moffett*   5

2 What Joe Loved, *Mark W. Moffett*   12

3 Skunk Oil, *Betsy L. Howell*   14

4 Connections Are Everything, *Ram Papish*   26

5 Creeping Death, *Howard Whiteman*   29

6 In Big Fence Country, *Michael Rogner*   38

7 No Forwarding Address for Vernal Pools, *Roy A. Woodward*   48

8 Running the Wind, *Jennifer Bové*   50

9 Swimmin' Hole, *Wendell R. Haag*   59

10 The Goat in the Galley, *Chris Smith*   60

11 A Fox's Love Story, *Ran Levy-Yamamori*   68

12 Fire: A Prairie's Own Companion, *Alice Cascorbi*   70

13 Road Hunter, *David M. Liberty*   76

14 Snares, *Troy Davis*   78

15 Pileated Woodpeckers, *Patrick Loafman*   84

16 Museum Sisters of Cheerful Disposition,
Barbara Blanchard DeWolfe   88

17 Up Trapper Creek, *Jennifer Bové*   90

18 Love in the Margin: Finding Refuge in the Blue Mountains,
Joseph L. Ebersole   97

19 The Clinch River Diner, *Charles F. Saylor* 104

20 The Science of Islands, *Scott Stollery* 107

21 Tigerland, *Eric Dinerstein* 110

22 A Mother's Worst Nightmare, *Tonya M. Haff* 132

23 Potheads, *Sharyn Hedrick* 136

24 Gumbo, *Lynn Sainsbury* 142

25 La Yunga, *Betsy L. Howell* 145

26 Rollie's Catch, *Chris Smith* 156

27 The Gift Eagle, *Dan Mulhern* 159

28 Tequila, *Patrick Loafman* 167

29 Polar Summer, *Cameron Walker* 172

30 Booby, *James Lazell* 179

31 Sandhill Season, *Jennifer Bové* 182

32 Eating Crow (and Other Ways to Atone for Sins in the Wilderness), *Chris Smith* 186

33 Chasing the Tail of a Frog, *Elizabeth Dayton* 189

34 Dances with Coyotes, *John Shivik* 201

35 Snapper's Kiss, *Jeff Beane* 206

36 Whistling Dixie in the Land of the Midnight Sun, *Benton Taylor* 208

37 Dependence Day, *Jeff Beane* 220

38 The Old Camp Coffee-Pot, *Badger Clark* 224

Acknowledgments 226

Contributors 227

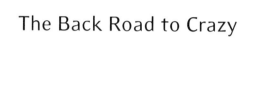

The Back Road to Crazy

Introduction

JENNIFER BOVÉ

Straddling a cedar the girth of a horse's back in the cold current of a remote southern Washington stream, I assessed the job before me. I had a ten-foot length of steel cable draped across my thighs that I needed to wrap in a strategic figure eight around the tree I was sitting on and another that lay submerged a good three feet beneath it. This connection of timber was just one of many that would secure a massive logjam against the creek's deeply eroded bank. The seemingly haphazard pile of large wood had, in fact, been thoughtfully engineered by Brian Bair, an intrepid fish biologist with the U.S. Forest Service, to buffer the wounded cutbank against the unrelenting force of flowing water. In the grand scheme, this effort would help restore the creek to a state of natural functionality that preceded riparian logging operations. It was a stepping-stone on the long, winding, and rugged path to salmon restoration in the Columbia River watershed.

But my task was straightforward that morning. I wasn't saving the world; I simply had to cable a couple of logs together.

I circled my feet in the water, feeling the icy sluice through my wading boots and jeans. It was just about eight-thirty in the morning, and the sun had not yet topped the Doug firs that shaded the creek. Goose bumps prickled my skin. I knew I needed to get into the water in order to thread the cable around the trees, but I was in no hurry to get wet. So I stalled. I adjusted the bulky cable clamps in my pockets. I looked for somebody—*anybody*—from my crew who might have finished working on the logjam upstream and decided to come lend me a hand.

I tugged the sleeves of my T-shirt down a little farther onto my arms and shivered.

It occurred to me right about then to wonder why exactly I was about to submit my body to the assault of mountain creek water at dawn. If I didn't suffer an immediate heart attack upon hitting the water, I knew I was guaranteed damp clothing and a stubborn chill in my bones for hours thereafter. There were, I was certain, a million other more comfortable ways to earn a modest paycheck.

This wasn't the first time I had asked myself "Why am I doing this?" in four years of working as a field biologist, and it would not be the last. In the name of conservation, I had navigated snake-infested sloughs reeking of cattle excrement, spent scorching summer afternoons in the woods scrubbing seed ticks from skin already inflamed by poison ivy, suffered bleeding blisters beneath wet boots for days at a time, and once waded neck-deep in the downstream wash of a very dead beaver—on each occasion questioning my own good sense.

The answer, I've found, has always been the same. In moments of reluctance and doubt regarding my chosen career, I remember a photograph I ripped from a magazine during my first year of college. In it a brawny female biologist carries a tranquilized wolf across her shoulders in a driving Montana snow. The image filled me with admiration, awe, and a sudden sense of direction in life. Looking at the photo, I could feel the winter, the wilderness, and the weight of the wolf on my back, and I understood that, for me, the burden of that animal was a metaphor for nature in need. I wanted to make a living like the wolf woman. I wanted to make a difference in one creature's life that might also impact the web of the world in some positive way. It was that photograph which called me into the wild and still reminds me, when my tolerance for outdoor "wonders" is stretched to the breaking point, that I was meant for this work. It's what I need to do.

Obviously, one length of steel cable in a series of mammoth log constructions along that Washington stream was far from a feat of environmental heroism. It was, however, a small and vital link in a chain of human endeavors that might draw Steelhead upriver to spawn again. With this in mind, I once more accepted a mission that felt a whole lot more like torture than a job probably should. I swung my leg over the tree, clutching the cable and a socket wrench in one hand, and I held my breath as I scooted off the splintery back of the cedar into the frigid creek to begin another day's work.

The Back Road to Crazy originated as a compilation of autobiographical stories based on my own experiences afield. I aspired to represent my nomadic, challenging, and often extraordinary profession in all its gritty reality. This book, I imagined, would offer a window to a living that few people are ever able to witness, much less undertake. I soon realized, though, that I couldn't do it on my own. Not only were there an infinite number of studies and places and species I had never encountered, but there were also so many different perspectives that I felt needed to be portrayed. It became clear that the only way I could convey the full scope of this occupation would be to offer a forum for the stories of other biologists who have worked in all varieties of field situations.

I sent out an international call for stories, to which my colleagues rallied with a remarkable array of writings from the field, and *The Back Road to Crazy* fledged into an unprecedented collection of creative nonfiction by biologists from around the world. Seasoned researchers and novices alike have come together in this volume to reveal the reasons they trade the comforts of more sheltered careers for physical labor, whims of weather, and the company of unruly creatures. These are not weekend naturalists who don khakis and cameras to explore the outdoors in their spare time. The contributors featured in this volume include state and federal biologists, professors, technicians, and students who have devoted substantial energy to pursuing decidedly unglamorous field studies. For them, the work is a way of life that hangs in delicate balance between sacrifice and reward. In their own words, these authors describe their individual impulses to step beyond the well-beaten pathways of parks and nature preserves, and they express a universal drive to dig their fingers into the gnarled and toothy earth in order to uncover secrets that might just save our precious wilderness.

Each contributor's account, whether humorous or tragic, illuminates a profession that pays its members to perform almost any kind of outdoor assignment imaginable, from traversing woodlands thick with grizzly bears to riding out furious arctic seas, all in the unyielding pursuit of scientific knowledge. Diverse in both subject matter and style, these stories collaborate fluently to represent biological fieldwork in a manner that is entertaining, informative, and intrepidly honest.

You will find a common thread of philosophy woven throughout the book's assorted writings. From hidden corners of America to distant reaches of the globe, biologists share the notion that in the course of fieldwork, what doesn't kill you will make you tougher, deepen your sense of ecological accountability, and provide you with a cache of memorable stories.

I

Bit

MARK W. MOFFETT

That morning I woke at dawn and crawled from my tent into the big unpainted schoolroom where the members of our biology expedition slept. We were in Rat Baw, a village in the far north of Myanmar. Outside, herpetologist Joe Slowinski and his best friend, photographer Dong Lin, stood wearing matching green T-shirts stenciled with one of Dong's photos of a cobra, poised to strike. I walked up as Joe's Burmese field team leader, U Htun Win, held out a snake bag. "I think it's a *Dinodon*," he was saying. Joe extended his right hand into the bag. When it reappeared, a pencil-thin, gray-banded snake swung from the base of his middle finger. "That's a fucking krait," Joe said. He pulled off the snake and kneaded the bitten area, seemingly unmarked, with a fingernail.

Other scientists have been known to cut off their hands at such a moment. Joe sat down to join the rest of us for breakfast at a long wooden school table, joking about his thick skin. It was seven o'clock in the morning on September 11, 2001.

I'd known Joe for two years, seeing him most often when he drove over to Berkeley for evening herpetology seminars at the University of California. A thirty-eight-year-old field biologist with the California Academy of Sciences in San Francisco, he had published papers on evolutionary theory, systematics, and the origins of biological diversity—but mostly he was the man to talk to about cobras. For ten years, Joe had been concentrating on the rich biological triangle of Southeast Asia where Myanmar—formerly known as Burma—and Laos meet southwestern China. He was conducting a comprehensive survey of the herpetofauna of Myanmar; on ten expeditions since 1997 he'd found eighteen new species of amphibians and reptiles, including a new spitting cobra, *Naja mandalayensis*—which he considered the "ultimate discovery." He planned to write the definitive book on Myanmar's natural history and hoped also to help establish a museum of biodiversity there.

Before a seminar Joe, Dong Lin, and I would share beers at La Val's Pizza. Dong, now in his mid-forties, told me how, after surviving Tiananmen Square with sixty stitches, he had escaped China in 1990 and made his way to a position in photography at Cal Academy. There Joe helped guide him through the book *English as a Second F**king Language*, and soon after, Dong started to join him as expedition photographer. Over our Coronas, Joe would describe his upcoming trips, slapping me on the back and telling his best adventure stories to entice me to "come along this time, bro."

As an entomology researcher at Berkeley, I recognized in Joe someone like myself, someone who in earliest childhood fell hard for a disrespected creature—in Joe's case, snakes; in mine, ants—and managed to retain that fascination into adulthood and even build it into a career. He had a boy's sandy hair and freckles, and his habitual expression of sheer uninhibited wonder was matched by a precise and agile mind. His fieldwork had the same old-fashioned sense of exploration I'd grown up admiring in nineteenth-century scientist-explorers like Charles Darwin and Alfred Wallace.

Time and again, Joe's schedule and mine had conflicted. Then one night in La Val's he described a trip coming up in September. He'd recruited colleagues from different disciplines to conduct a broad species inventory of Burma's remote northern mountains. Perfect.

The expedition would take us into the foothills of the Himalayas; it was scheduled to last six weeks and span two hundred miles. Our group of eight American and two Chinese scientists, joined by four Burmese research assistants, gathered on September 3 in the village of Machan Baw—the dusty remnant of an old British outpost—and started walking, accompanied by a long line of porters. Machan Baw sits at 1,400 feet; the plan was to climb above 10,000 feet, surveying a range of habitats from subtropical forests to temperate highlands and traveling eventually into the new Hkakabo Razi National Park.

Adventures are made mostly in the recollecting mind; the doing is generally more drudgery than drama. It was monsoon season, and our path, more mud trough than trail, was hard slogging. Leeches emerged in droves. We tried to keep them at bay by spitting tobacco juice onto our legs or wearing panty hose, but Joe, trekking in shorts and sandals, simply put up with them, as did many of the porters. At times I'd look down and see that the rain puddles along our route were red with blood.

The first week took us through farmland and villages. Houses with roughly stacked pole walls were raised on stilts so that pigs and chickens (and their legions of fleas) could sleep in the slightly protected muck below. Each evening sand flies speckled our arms with welts while mosquitoes threatened us with a malaria resistant to most prophylactics—one reason we zipped ourselves into tents even when sleeping under a roof.

In patches of rain forest between rice paddies we found enough species to keep us moving eagerly toward richer territory. The sonic duet of gibbons and two huge-beaked hornbills passing overhead indicated more pristine habitat nearby. After each trek Joe would gather bags with the day's specimens from his Burmese team and from our frog expert, Guin Wogan, a graduate student at Cal Academy. Dong Lin would video the most unusual individuals. If venomous snakes were involved, Joe would wrangle them so Dong could get the best footage, shooting from inches away—greatly impressing the inevitable crowd of Burmese onlookers.

Joe was careful with snakes. He was also famous for his close calls. Bitten by a copperhead as a child in Kansas, he'd gone back the next day to catch another, left-handed. On a previous trip to Burma a spitting cobra had struck through the bag Joe had put it in, stabbing his finger. He waited calmly for the venom to take effect. Luck of the draw, he would say, telling the story: sometimes a snake bites without injecting its toxins. On a later Burma trip a cobra squirted venom into his eyes. After a few hours the excruciating pain passed. Joe never paused much over these incidents. He seemed to embody the understanding that a fully natural world includes the possibility that nature can kill us—and afterward glide freely away into the wet grass it came from. That love in any form involves an element of risk.

It was good to see Joe at work in the country he'd described so often. He was proud of his Burmese field assistants, on permanent loan to him from the Myanmar Forestry Department. In a country with few scientists, Joe saw these young men and women as an essential resource for the future. Species inventories are a big part of conservation, and his assistants caught, preserved, and documented specimens year-round. Joe had struggled hard over the past five years to build government contacts; research in heavily militarized Burma is no simple thing.

Returning late at night by headlamp, Joe would unload his catch of snakes and frogs and sit with whoever was still awake, usually Dong Lin and me. During our conversations I began to see the different sides of my friend. Some nights it seemed he felt invincible. Downing Burmese rum, he knew he would rise high enough in the hierarchy of science to put a stop to the "political bullshit" he saw all around him. Much of what he imagined seemed possible: he'd just been awarded a $2.4 million grant from the National Science Foundation, already a sponsor of his ongoing Burma research, to study biodiversity in China's Yunnan province. He confided a thousand ambitions, certain he'd realize them all.

Other times Joe raged into the night, once about another biologist working in Burma who Joe believed had blocked the original funding for this trip. Joe had hastily cobbled together funds from his other grants and gone anyway. His tirade explained something. I'd wondered why our expedition had come during the rainy season, when (as was evident once we started walking) we could have taken Jeeps along much of the route any other time of the year. Remembering how discovery breeds rivalry and how scientists can turn research into races, I sat in a small dry spot surrounded by what seemed a world of mud, an understanding comrade to Joe's fury.

Still other nights Joe grew melancholy. For years he'd focused only on science, he said; he'd been too single-minded, traveled too much, even for love. Now, though, he'd started a relationship with an ornithologist back home. He wondered if he should devote less time to snakes.

Managing the people and logistics along with his research on this trip was clearly taking a toll. There was a lot to worry about. Among the multitudinous supplies we'd brought were drying ovens and pounds of newspaper for the plant specimens, snap traps and mist nets for the mammals and birds, gallons of alcohol to preserve reptiles and insects, a generator and its gallons of fuel to recharge batteries for cameras and computers and to run the black light for attracting insects.

Ninety-odd porters hauled the equipment of ten academics. Many of the inevitable problems were handled by a Burmese guide, but Joe had to think about them all. In addition, he'd paid $44,000 to a well-connected expedition coordinator to cover the in-country expenses, yet somehow such basics as rice and bottled water were in astonishingly short supply, so Joe kept spending more, out of pocket. Nor was there any sign of the two military doctors and radiophone the government had promised. Joe guessed the real cost of the trip was probably a tenth of what the expedition had put down.

Then there were the scientists. Each of us wanted to work at our own pace and had our own agenda. Personalities often clash in the field, and for Joe, feeling responsible for the group's harmony must have been one more stress, along with our long daytime treks and his own additional nocturnal collecting. I noticed the accumulating effect on him during a walk on September 10, only the seventh day of the trip. He was moving sluggishly, and each time he paused to pull a leech from his leg, his fingers were visibly shaking.

After it was over, we'd all wonder why Joe had reached into the snake bag with barely a glance inside. As with any pivotal moment, the exact words exchanged beforehand would be endlessly debated. Snakes of the genus *Dinodon* are harmless, but some are near-perfect mimics of the Multibanded Krait (*Bungarus multicinctus*), a cousin of the cobra and much more deadly. U Htun Win should have known the difference—he told us he'd been bitten by the snake the day before and nothing had happened. But Joe was the authority. Possibly simple exhaustion brought his guard down; perhaps he failed to heed the uncertainty in U Htun Win's tentative identification.

Following breakfast, around seven-thirty, Joe lay down. At eight o'clock he noticed a tingling in the muscles of his hand and asked Dong Lin to call the group together. By eight-fifteen two Burmese assistants started the run of eight miles to Naung-Mon, the nearest town with a radio. Joe calmly told us what would probably happen and what we should do. He described the effects of a slowly increasing paralysis, eventually requiring mouth-to-mouth respiration until he could be taken to a hospital. If he lived, the neurotoxins would work their way out of his system in forty-eight hours. He would be conscious, he told us, the whole time.

As the morning went on, Joe had to reach up to open his eyelids. His breathing grew raspy; his voice was reduced to a slur. In time he could only write messages: "Please support my head—it's hard for me." "If I vomit, it could be bad." "Can I lean back a little." By noon he could no longer breathe on his own. "Blow harder," he wrote. In his final message, minutes later, Joe spelled out "let me di."

"We won't let that happen," Guin Wogan said.

"Kick butt, Joe," I added.

At three o'clock our runners returned alone—the military had requested an update before it would send an air rescue. Two fresh assistants were sent back, again insisting that a helicopter be sent. By evening the weather turned from the best we'd seen in a week on the trail to a renewed downpour; low clouds would impede the rescue again the next day. That night soldiers arrived on foot with an ancient field radio and a young Burmese doctor with two nurses and a little equipment, including an old respirator no one could get to work.

Throughout that long night we all helped out as we could, but much of the time was spent in simple exhausted witness. From time to time Dong would put his arm around various members of the group and say "I love you." In one long moment of vertigo, as someone who's had his own close calls with snakes, I looked at Joe in the torchlight and saw how alike we were in build, complexion, even our features, and I felt I was somehow watching myself die. Looking at Dong, Guin, and U Htun Win standing silently nearby, I wondered if they felt something similar.

By three in the morning Joe could no longer signal us except with his big toe. His final communication occurred when ornithology assistant Maureen Flannery, whose strength had been keeping us all going, asked if she and Guin could stop doing mouth-to-mouth and let the guys take over. Joe's toe signals indicated a preference for the women.

During the twenty-six stifling, sand fly–infested hours that the artificial respiration continued, four airliners plowed into their final destinations in New York, Washington, D.C., and Pennsylvania. The only one of us who knew was David Catania, a Cal Academy ichthyologist so understated that I often forgot he was there. Dave had listened to his shortwave radio after collapsing briefly in his tent late in the night. Keeping the news to himself, he came out and gave Joe mouth-to-mouth for hours, his face showering sweat. He refused to let anyone else take over, even long after Joe's heart had stopped.

At twenty-five minutes after noon on September 12 the doctor told us Joe's pulse was gone. We began three hours of CPR in anticipation of a rescue helicopter that was never able to land.

Joe's body was cremated in a small Buddhist ceremony two days later in the town of Myitkyina, and Dong Lin and some of the team brought his ashes back to San Francisco, along with many of the expedition specimens. The others made their way home as best they could.

It was not until two months after Joe's death that I returned from Asia and made my way to ground zero in New York. Even faced with the physical evidence of the events in my own country on September 11, evidence from which white smoke still rose, the devastation seemed impossible to take in, the tragedy too strange and vast for my already reeling heart and mind. The unfolding of Joe's death in the same hours had been so intimate, so personal, and at the same time so dramatically dull.

For everyone in Rat Baw but our team, September 11 seemed to have been an ordinary day. No one had heard of the events in distant America, and Rat Baw was a place where death from such natural causes as snakebite was a common event: there are more deaths from snakes in Burma than anywhere else in the world. Children played in the field within yards of the room where Joe lay among our small circle giving him CPR. Elders sat on benches outside, hour after hour, talking softly and watching the rain, as one supposes they always had.

One of Joe's gifts was the way that for him the ordinary always seemed to yield to the extraordinary. I thought more about this later, when my mind had finally begun to move beyond the trauma of his death. I suddenly recalled something that had happened the day before the bite. Joe had returned from a walk in Rat Baw flushed with excitement: he'd found a pair of entwined kraits. "It was beautiful. Goddamn beau-ti-ful! Courting like that, right in the middle of the trail. I've never seen anything like it." His arms sliced arcs in the soupy air. The weight of all our petty concerns had vanished from his face, and his eyes seemed to glow, as they always did at moments like this, with the love of snakes.

2

What Joe Loved

MARK W. MOFFETT

Last night, dreaming,
I saw again that thin, gray-banded snake
in ropey grass.
A krait, inches-long cousin to the cobra.

Its heart beat a serpent's rhythm.
Long and measured. Slow and careful.
Like the pulse in the wrist of a dying friend.
Who is to say which deserves to live,
the friend because he is a friend,
the snake because its movements
whisper perfect through the dew-bent grass?
Who can say which one's heart beats a greater truth?
My friend's because he loves the snake.
The snake's because it fears the man.

My friend carries a syringe of formaline
because loving snakes means killing them for study.
The snake carries tiny, gleaming teeth
glazed with a different poison
because fearing men means killing them to live.

It might take the krait its lifetime
to make one poison drop
the size of the tear that sits in my friend's
unblinking eye.
Kneeling over Joe, I hear his heart.
Still drumming a man's rhythm, yes,
but indistinguishable now from the snake's
as it fades
like a whisper slipped into wet grass.

The gray bands flash and vanish.
The squatting tear evaporates.
A love past fear plays out its final beat.

3

skunk Oil

BETSY L. HOWELL

I picked up the ringing phone. Nancy, our district purchasing agent, was on the line.

"A package just arrived for you," she said with a nasal tone to her voice.

"Great. I'll be up in a few minutes."

"No," Nancy replied firmly. "Come now."

I walked across the compound of the Forest Service office where I worked as the district wildlife biologist. The autumn air was as clean as a baby's mind. Fall has always been my favorite time of year. The desiccated leaves of summer, contorted into a million shapes, the mix of cold air and warm sunshine, and the darkening days of winter remind me of earlier times when humans were not so separate from nature. In this valley, where I now lived, home to the small town of Powers, Oregon, you could smell the smoke from wood stoves against the biting cold. Gunshots rang out from hunters, who chased after Black-tailed Deer in the surrounding hills.

As I walked into the front office, the odor of skunk assailed me like nothing I had ever experienced driving by a roadkill. *Oh-oh*, I thought, *IT has arrived.*

Nancy and the other women in business administration did not look happy. Donna, the receptionist, held a bandanna over the lower part of her face. Jodi, our computer specialist, frowned as I walked by, her nose pinched between two fingers.

"I guess my skunk oil has arrived, eh?" I tried to joke.

"Yes," said Nancy, pointing to a box sealed with clear, sturdy tape.

"The UPS man's face was green as he delivered it. Said the eighteen-mile, winding drive up here, with your package, almost made him stop the truck and retch. Here," she handed me the packing order, "sign your John Hancock and take it."

"Shouldn't I make sure it's what I ordered?" Nancy always made me open my purchases and verify the contents. Now she looked chagrined at the prospect.

"If you want to do that, go outside."

"Way outside!" someone gasped from the hallway.

I took a big gulp of air and held my breath as I carried the box to the parking lot. The return address read, *M & M Fur Company, Bridgewater, South Dakota*. A small crowd gathered at a distance. My supervisor, Robin, braved the smell and helped me open the package. Inside, buried beneath wood chips, was a paint can. From inside the paint can, among more wood chips, I pulled out four five-ounce bottles of pure skunk oil, straight from the anal glands of this continent's most odiferous creature. Just what I'd ordered.

The previous summer the Forest Service had bought six remote infrared cameras to begin a winter survey for American Marten and fisher. Both species were presumed to occur on our district, the northern portion of the Siskiyou National Forest in southwest Oregon, but we had only anecdotal sightings as evidence. At the time I began this study the marten had been designated a "management indicator species," meaning it represented a host of other species that use old-growth coniferous forests. Theoretically, providing adequate habitat for martens would allow for the perpetuation of populations of other forest dwellers as well.

Unfortunately, in the mid-1990s little was known about either species in this part of their range. These mustelids, members of the weasel family, are notoriously difficult to document. They are nocturnal and secretive, do not leave behind scrapes or scratches, occur at low densities, and avoid humans whenever possible. Capturing individuals on film was the most cost-effective, as well as least intrusive, method to document martens and fishers on our district. Additionally, I knew it would be fun. Several years before, I had worked with "line-trigger" camera set-ups, which used inexpensive 110 cameras tied with a string to a small piece of chicken. When the bait was tugged, a bent wire would depress the shutter button. With these systems I got pictures of bears, spotted skunks, a Long-tailed Weasel, and one wood rat. It was like Christmas every time I got a roll of film developed; I never knew what I'd find. However, only one picture could be taken with the line-trigger method before I would have to return and advance the film. With the automatic advance on the infrared cameras, an entire roll could be exposed.

I transported the skunk oil to the South Compound, our Forest Service storage buildings, approximately two miles from the main office. Into a garbage can I placed the box, securing the lid tightly. My Chevy S10 Forest Service truck stank the rest of the day. How I was going to transport the oil to the survey locations without becoming sick myself wasn't yet clear.

A few days later Tina, the supervisor of business administration, called me into her office.

"Betsy, what's in the bunkhouse freezer?"

"Why do you ask?"

"We have some guests staying overnight, and when I showed them the fridge, *and* freezer, I saw it was packed with plastic grocery bags filled with... well, something."

"Those are bags of bait for the camera traps," I explained. "They won't be there for long. I used up all my own freezer space and had nowhere else to put the, uh, stuff."

I didn't think Tina wanted to know what the "bait" was. I had obtained permission from the Oregon Department of Fish and Wildlife to collect roadkill deer for the project. Driving back to Powers late one night the week before, along the South Fork of the Coquille River, I came across a recently hit doe. Never having been a hunter, I fumbled in the dark for some time employing ropes and what I could remember of physics to hoist the animal into the back of my Subaru wagon. Later a friend helped me cut off the legs, and these I put into the freezer of the boys' bunkhouse.

The ODFW also generously donated spawned salmon from the local fish hatchery and beaver that had been trapped on animal damage complaints. Dave, another biologist on the district, told me I would get interesting results with beaver bait.

"Wildlife really like the taste of beaver," he said. "The castor is great for using on traplines. And everything will investigate the skunk smell."

The bait I would hang from a tree with one-eighth-inch wire cable, the camera facing it so as to photograph the animal as it went for the treat. The infrared beam would cross underneath the bait, and the camera, attached to the beam's transmitter unit by another cable, would go off just as the animal was reaching the salmon, or deer, or beaver. Using an eyedropper, as instructed by the camera survey protocol, I would place a few drops of skunk oil on a feminine maxi pad, poke a piece of wire through it, and enclose the pad in the top half of a gallon milk jug cut in two. This way the pad would stay dry during the wet Oregon winter and exude "eau de skunk" to attract visitors.

My learning curve that first winter was steep and exhausting. The Siskiyou Mountains are precipitous country, tangled in places with salal and rhododendron that can grow twice the height of a person. I have walked through woods where my feet never touched the ground, relying for footholds on the rhododendron trunks. Three large rivers flowed on the district, the Elk, South Fork of the Coquille, and the Sixes. Into each of them cascaded hundreds of smaller streams. It was a landscape that had challenged engineers over the decades as they determined the best way to road the slopes and ridgetops. The topography, with its incised canyons and numerous waterways, allowed only crooked, meandering passage. It often took me two hours to drive thirty air miles.

Because I had limited time for camera surveys that first winter, I chose to put the stations close to our district office. I placed them at elevations lower than that typically occupied by martens and fishers because I did not have access to a snowmobile. For this reason, I obtained pictures of a lot of wildlife that winter, though none of the elusive species I wanted.

The day before the project began, I drove to the ODFW hatchery on the Oregon coast, where they had a large walk-in freezer. I collected four beaver, one of which weighed fifty pounds. None had been cut up into manageable chunks. With no freezer space available, the beaver sat in my truck overnight. The next day I headed to the woods with the carcasses in burlap bags, the skunk oil in its paint can and box, four empty milk jugs, a box of maxi pads, and the camera equipment. Robin had raised his eyebrows when I submitted a requisition to buy the Kotex.

"It's for the camera project," I explained, embarrassed. He nodded without smiling.

I set up the first station in the Douglas fir woods around a small pond known as Azalea Lake. In the spring the steep hillsides were carpeted with blooming azaleas, as if someone had dolloped scoops of fragrant whipped cream beneath the forest canopy. In October, however, there were no tangy springtime smells, only the pungent aroma of skunk and beaver that had been melting since the previous day. I parked near an old logging road, which I would use to access the site. Steller's Jays came immediately to my truck to investigate the smell. With the summer birds gone, I heard only the nuthatches with their annoyed *eh-eh-eh* call; the high-pitched voice of the Golden-crowned Kinglet, which reminded me of someone leaping high off a diving board, then dropping into several somersaults; and the Varied Thrush's dirge, which sounded like a telephone ringing in the forest.

The camera survey protocol recommended having two people set up the stations, one to handle the smelly bait and skunk essence, the other to arrange the camera and infrared units. The authors warned that getting the animal smells on the equipment could result in damage if a visitor, particularly a bear, became too engrossed in the technology. They also advised making sure winter was in full force, and bears dormant, or at least less active, before setting the stations up. But southern Oregon winters never get really cold. I didn't want to wait.

The camera equipment, along with a small saw, duct tape, station tags, a staple gun, and flagging, was stowed in my backpack. I pulled the biggest beaver out of the truck and decided it didn't weigh that much. Now for the skunk oil. I donned my rubber gloves, tied a bandanna tightly around my face, and took a big breath of air. In less than a minute, when the air would run out, I had to get the paint can out of the box, open it, fish out one of the bottles, open *it*, get an eyedropper full of the brown liquid, squirt it onto the maxi pad, and seal the bottle back in its woodchip tomb. I never made it. It was too much to do without breathing. I had to stop, run thirty feet from the truck, breathe for several seconds, then try again. The problem with skunk oil at that concentration and proximity is that it gets into your nose and mouth even when you're not breathing. Even when you have your mouth closed and covered with a bandanna. I dry-heaved the taste and smell out of my mouth. My eyes watered. I spit on the Oregon grape and evergreen huckleberry bushes and gasped. There had to be an easier way, but I didn't know what it was.

Each station required three trees, at least six inches in diameter, in a triangular configuration. The infrared boxes would be tied on two of the trees, the camera on the third. Still wearing my bandanna, I headed up the steep logging road, dragging both the maxi pad, tucked in its milk jug, and the soft, dead weight of the beaver. I hadn't gone but thirty yards and had to stop. Sweat deluged my eyes and saturated the scarf. Fifty pounds of beaver was a lot of beaver. I panicked. If I couldn't haul this on the ground, how would I ever hoist it into the air? The bait needed to be up high, attached firmly to the tree, or it would likely be stolen by whoever got there first. I set everything down and rested farther up the trail away from the smells. I was disgusted with myself for turning into an "office biologist." *Well, that will change this winter*, I thought.

A half-mile up the road I found three trees I thought would work and began setting up the station. The infrared units needed to be lined up exactly, the camera securely affixed to the tree, and the cable linking them strung high above so it wouldn't be torn down. Everything went smoothly until I came to the beaver. I poked the wire cable through its flat, scaly tail and threw the other end over a branch. As I hoisted, the branch broke. I tried another. It broke too. A third proved strong enough, but then I wasn't. That beaver wasn't going anywhere but on the ground. I finally just tied it to the base of the tree and hoped for the best.

I set up the rest of the stations that winter at places known as Johnson Mountain, Rock Creek, and Big Tree. The salmon and deer legs I hung easily in the trees; the beaver I secured on the ground. I became more adept at moving quickly when preparing the "skunk pads," but I still gagged every time. And despite the fact that I left the skunk paint can at the South Compound, my truck took on a permanent odor of its own. Robin approached me one day about it.

"There have been some complaints around the office about the, uh, stink of your rig."

"Really?" I was surprised. After being up close and personal with a bottle of pure skunk oil, this lingering bouquet, which Robin referred to as "stink," did not seem so bad.

"Yes, a motion has been made that you move your truck over by the bunkhouses," he said. I felt ostracized but also enjoyed the notoriety. My co-workers already thought I was strange with my skull and scat collections. Now their worst fears were coming true. My interests had begun to affect everyone. Tina continued to query me about when the bunkhouse freezers would be available again. Though she did this with an amused countenance, she was not entertained when another district-wide complaint came to her attention.

"It seems there are 'mystery bags' in the freezer of the district fridge now as well," Tina said. Since this community lunch fridge was right next to her office, I was sure she had heard several overblown stories.

"I've run out of freezer space again," I explained. "There are only a few pieces of deer and salmon in there." To lighten the situation, I added, "At least it's not skunk."

"Maybe you need to get your own freezer," she suggested.

When I returned to the Rock Creek station, also placed along an old logging road, I found my work destroyed. The transmitter lay on the ground, and its red light, which indicated a properly functioning beam, was glowing eerily. Holes the size of inch-long canines speckled the unit, and scratch marks had torn up the duct tape I'd used to adhere a small metal roof above the camera. My impatience to begin the project had paid off in a damaged system. Obviously, a bear had come by and decided a free meal wasn't enough. I packed up everything, disappointed.

Each station remained in the woods for one month. During that time I checked every week to replace film and batteries, hang more bait and skunk scent, and confirm that all was operational. I took notes on what I observed: holes dug in the ground and equipment that had been moved or chewed. I had little hope for martens and fishers since they preferred higher elevations, but I considered this a learning season. As the first rolls were developed, I was excited to see many "captured" species. Spotted skunks visited frequently and would "camp" at a location, using up all the film. Bobcats also were not bothered by the camera's flash and motor's whirr, and I would have twenty-four shots of one animal in various poses. Black-tailed Deer, Steller's Jays, and raccoons came regularly.

It rained a lot that winter, typical for southwest Oregon, and my raingear became saturated with the smells of dead salmon and beaver. I could not avoid slopping entrails, blood, and scales over the nylon fabric. Washing did not help. My co-workers told me outright that I smelled. Consequently, I received fewer interruptions during my office days, especially if I left my raingear out to dry around my desk. But as they say, "one species' bane is another species' treasure." Where humans had no appreciation for this rich variety of fragrances, other mammals did.

In November I attended a carnivore workshop on the snowy slopes of the Mount Hood National Forest. Speakers came from several western states to discuss four carnivores that represented endangered old-growth coniferous forests: American Marten, fisher, wolverine, and Canada Lynx. There was lively debate about inventory methods, including camera systems, track plates (sooted with black smoke to register an animal's footprint), and snow surveys. A man named John Weston from Montana brought his tame lynx, Chirp, to demonstrate what the tracks of an animal look like under different forms of locomotion, that is, running, trotting, and walking.

One morning we went to the woods around Timberline Lodge to watch Chirp in action. Though the snow had piled up over the previous weeks, only the steady drizzle of rain greeted us. I rode with John to a snow park on the east side of Mount Hood. Chirp sat in the back of the truck like a golden retriever excited about her next walk.

"How did you find Chirp?" I asked, marveling at the cat's perfect form and intent look. From my earliest memories, cats, especially Mountain Lions, have been the treasure of my existence. When all else seems bleak or eroding in my life, I think of the matchless wonder of the cat, wild or domestic, and I know that at least one thing makes sense in the world. This animal can move without making sound, jump almost without limit (Mountain Lions have been known to leap down from heights of sixty feet), and hunt with a precision that makes them deadly to creatures as small as shrews and as large as elk. When I started the camera project, I had just returned from working in Argentina on a puma project, a fulfillment of a dream to work with wild cats.

John was evasive about Chirp's origins. "The important thing is *why* I got Chirp," he said. "She's what I call an 'ambassador animal.' We travel to schools and other groups and educate people about wildlife conservation, the importance of habitat preservation, and simply knowing about the other species that live with us."

In the woods we watched Chirp make her way through the wet snow. John let her run free, then brought her back to the group with a long leash. I sat on my knees. Chirp came my way and found with much delight that I smelled like a skunky, fishy, rotten-flesh cocktail. She licked my rain pants and coat and rubbed her cheeks and face over my thighs and stomach. It felt like heaven to be so close to a wild animal, even if the circumstances were artificial. Soon Chirp was so excited, much like my housecats with catnip, that she began to gnaw on my gloved hand. She could have eaten the hand—I didn't care.

Later that winter I checked a station on Johnson Mountain. At this site I had tied a salmon in a tree and left a beaver, with the cable through its tail, on the ground. The skunk jug dangled against the dripping fish. Most of the camera stations, even if the film was used up, held few clues as to the visitor. A small hole dug in the ground might be a skunk. Total demolition would be a bear. But often I wouldn't know until I got the film back. This day, however, I knew.

I stood amazed, looking at the cable that held the beaver. Aside from the tail and part of the back, which had been chewed on, the bait was now completely covered. Branches, leaves, and dirt had been scraped over the body. Mountain Lion, an animal so rarely seen it could be considered a ghost, had come to investigate and appropriate my gift for itself. This surprised me because I had read that lions would not scavenge, preferring to kill their own food. This trait is one of the reasons they were not wiped out with wolves and grizzlies during the predator exterminations of the early twentieth century; they would not eat the poisoned bait. Obviously, the "land beaver" I left had been too tempting to pass up.

I spent longer than usual at this site, making notes, taking pictures, and moving the camera closer to the bait. Walking in concentric circles around the station, I scanned the forest for other evidence of the cat but found nothing. I looked hard through the dark, dense trees. If wishing were enough, I would have seen the animal, which I felt sure was still in the neighborhood. Mountain Lions make kills, eat and cover them, then usually stay nearby, feeding repeatedly in the ensuing days. I knew the cat was watching and waiting.

When I got the roll of film back from Johnson Mountain, I found two photos of a Mountain Lion. The first, taken on December 5, shows an adult cat on the other side of the trail, its eyes glowing in the washed-out light of early morning. The second, from December 7, is a close-up. The cat is looking intently at the bait with a front paw raised. The ridges of muscle look like ocean waves, the thick, yellow tail disappearing into the forest shadows. I had checked the station on December 6. Now I was certain that the cat had been watching me while I moved the camera and made my notes. Though I wanted to find martens and fishers, I felt elated capturing a Mountain Lion on film. Showing the pictures to everyone, I explained my contentment knowing the cat had been watching me. "You're strange," they said.

The following camera season I bought a freezer and put it in the fire warehouse. The salmon and deer fit fine, but the adult beaver was still too big. Fortunately, I received a call one morning from a local trapper.

"I heard that you're collecting beaver for a wildlife project," said Kirk Hopp. "I have several in my freezer you can have."

I met Kirk in the nearby town of Coquille, and we hefted the beaver into my truck. When I told him of my storage dilemma with these whole animals, he had a suggestion.

"Cut them up into quarters. That's what I do for my trapline. Just take an ax, while the beaver is still frozen, and cut it in four sections. It's easy and not messy."

Messy is a relative term, and one a trapper might define differently from a gardener. Later that day I adroitly borrowed an ax from the district's tool supply and hauled the beaver behind the fire warehouse. I wouldn't have called the task clean or easy. I wore my fetid raingear and chopped beaver. Hoping no one saw me from the nearby highway, looking like a madwoman with an ax, I found Kirk's process to be slow but effective. I had a pile of beaver parts when Dorothy, the grounds-keeper, walked up.

Dorothy was an older woman who tended the Forest Service lawns. She was part of a large local family, rumored to have few branches in its ancestral tree. Stout, with a long braid of white hair down her back, Dorothy had always made me a little nervous. She was unpredictable. I'd walk by her and say hello, and she'd stare ahead as if she'd never seen me. Other times I'd ignore her, and she'd greet me. I had hoped to finish my ghoulish task before Dorothy found me. She stood out of range of the frozen flying intestines without speaking.

"Looks like you're tearing up the grass," she finally said, as if she observed someone cutting up frozen animals on the compound every day.

That winter I determined to put the stations up on the high peaks of the district: Mount Bolivar, Salmon Mountain, and Hanging Rock. It didn't snow a lot, but to reach those areas I needed four-wheel drive. One day I borrowed my co-worker's truck, accidentally leaving a piece of cotton saturated with skunk oil in the bed. Instead of the maxi pad system, I had decided to change to a smaller version. I soaked one piece of cotton, then ran a wire through it and the bottom of a film canister. The canister kept the cotton dry and was more manageable than a half a milk jug and maxi pad. It still reeked, however. For several days, the compound smelled of skunk yet I proclaimed innocence, as my vehicle was parked far away. When my oversight was discovered, Mac, the fire foreman, doused Guadalupe's truck with water and chemicals. Later I found this note on my desk: *Bets: Steam-cleaning was about 90% successful, i.e. for 100% deodorization, I would suggest immersing the entire rig in a vat of tomato juice & leave it overnight—Mac.*

It wasn't many weeks later when Mac paid me another visit. He looked disgusted, in a good-natured way.

"I need to talk to you about your freezer," he began. "It seems *somebody* left the door open overnight." Mac raised an eyebrow. "I walked into the warehouse this morning to find blood all over the floor. I'm sure you'll take care of this right away." And I did, but I wondered if the district was losing patience with me.

I worked hard from November through March, carrying beaver, camera equipment, and skunk cotton through the snow and up the mountains. But there still were no photos of martens or fishers. No Mountain Lions either. To be out alone in the serene, snow-laden forest was a gift. The Siskiyous calmed my mind at a time when other parts of my life were falling apart. My mother had died in October, and my only solace that long, dark winter was getting into the woods where the cold air neutralized the pain of feeling abandoned. But I still hoped for documentation of the species that were proving even more mysterious than a Mountain Lion.

Disappointed, at the end of March I packed up all the camera stations and put the equipment in storage. Two more rolls of film needed developing from the Salmon Mountain station. That week, at a district meeting, I won the Rusty Hinge Award, given for valiant, if comical, attempts to carry out one's duty. My co-worker Joe even penned a poem for me called "Dances with Skunks," which he read for all to hear.

I beamed that I'd had such an effect on everyone. Robin whispered to me as they clapped, "We'll never forget you."

The next week I picked up the film. Frame after frame revealed skunks and raccoons. My heart sank. As improbable as it sounds, however, when I came to the last slide, I was shocked to find a marten peering back at me from the bait tree.

Another forest phantom had been found.

4

Connections Are Everything

RAM PAPISH

Snake chaps, under the best circumstances, do little to flatter anyone who wears them. And although comparatively handsome chaps made of fang-proof fabric are available today, these options were not available to me back in the mid-1990s at Laguna Atascosa National Wildlife Refuge. Required field gear instead included plastic army-surplus units that protected me from possible run-ins with rattlesnakes but made me look like I had prosthetic legs. Not the type of thing you would wear to a job interview, to say the least.

In field biology, connections are everything. So I was very pleased to hear that the top people from the Peregrine Fund had flown down to Texas to see how things were going with our endangered Aplomado Falcons on the refuge. It was a fine day in early spring when they arrived, and though I missed them at the office that morning, I figured I would run into them in the field later that day because I needed to go out and check the falcons' artificial nest platforms. This would be the perfect opportunity to drop a hint that I would be interested in working for them that summer. Of course, they would understand about the snake chaps.

After a surprisingly cold winter I was delighted to be wearing shorts under my snake chaps on this warm spring day. My big blue shorts. I love those shorts. A little unprofessional, perhaps, but the Peregrine Fund people, they would understand.

I headed out into the field to check the platforms for evidence of nesting activity. The falcon prefers very open short-grass prairie punctuated by the occasional bush or yucca tree. The platforms are located high in these isolated yuccas, strung out along a quiet side road that is closed to the public. Checking them is simply a matter driving my truck along the road, keeping an eye on the odometer, then stopping and strolling through open grasslands at the correct compass bearing until I found the nest. I carried a long pole with a hubcap attached to one end that looked something like a giant dentist's mirror. Holding the hubcap end of the pole above a platform, I could gaze at the reflection in the hubcap and determine if falcons were nesting there that year. The hubcap-ended pole was, I realize, a curious-looking item to carry around, but it could easily have been explained to a fellow biologist.

On one of these truck-to-platform jaunts, feeling delighted with the weather, I took off my shirt as I walked and tucked it into my big blue shorts. When I returned to the truck, I threw the shirt onto the seat beside me and proceeded to drive down the road to the next stopping point. The next time I got out, I looked at the shirt there on the seat and thought, "Who needs it?" I straightened my ball cap, adjusted those stiff snake chaps, and set out at the proper compass bearing. After walking for about a half hour, enjoying the sunshine and gentle breeze, I realized that I had gone much too far. Looking back, I spotted a likely patch of yucca far in the distance. Then, looking at my shoulders, I saw that they were already turning quite pink, and I knew I was headed for one hell of a sunburn.

The solution was obvious. I removed my big blue shorts. Using my belt, I hung them cape-like around my neck, protecting my delicate shoulders and back from the cruel rays of the sun.

Picture this heroic figure walking through the open plains: Ball cap. Blue cape. Bare-skinned, hairy humanity. Staff crowned with hubcap. Tightie-whities. And plastic snake chaps.

I was quite near the road when the Peregrine Fund people, driving by, slowed their truck to a crawl.

I looked at them.

They looked at me.

I looked at me.

And I ran.

I ran directly at them.

I had to explain! They would understand!

But by the time I reached the road, the Peregrine Fund people were long gone. I just stood panting in their truck's skid marks as I watched my chance at making a great connection disappear down the road in a cloud of dust.

5

Creeping Death

HOWARD WHITEMAN

So let it be written
So let it be done
To kill the first born pharaoh son
I'm creeping death

—J. HETFIELD, L. ULRICH, C. BURTON, AND K. HAMMETT, 1984

The hike was long and tiring, but mostly downhill, and our pace quickened as we saw the Toyota truck through the Engelmann spruce. My assistant and I had been living near a group of high-elevation ponds in the West Elk Mountains of Colorado, studying the breeding behavior of Tiger Salamanders for the past few days as a part of my graduate research. We were tired, sick of ramen noodles, and ready for a hot meal and a shower at the Rocky Mountain Biological Laboratory (RMBL), just a short drive from where the truck was parked.

We loaded our packs in the truck, jumped in, and I turned the key. Nothing. I tried again. Still nothing. I knew all too well what the problem was likely to be. Jumping out of the car, I lifted the hood to inspect the damage. Miscellaneous wires, which I had meticulously fixed just days before, had been cut. Rubber hoses had been gnawed on, and claw marks decorated the inside of the engine well and the underside of the hood. Marmots. "Not again!" exclaimed my harried assistant.

I should have known better. Actually, I did know better. The marmots had been hitting me for years in this spot, or close by. But I needed to do my research, and this was the best place to park the car, even if it was right on top of a rock outcrop used by a large marmot colony. Once

29

the snow plug was gone, I could drive closer to the ponds and park in an area relatively devoid of marmots. For the better part of my field season, however, I would have to park here and suffer the consequences. Or join the battle.

After fixing the necessary wires with electrical tape (always have a roll handy in marmot country), I managed to get the car started and head down the hill. By this time we were late for dinner, but I knew we could put together some leftovers. I was just happy the marmots hadn't tampered with the brakes. As it always did, my mind worked the problem of how to get rid of the marmots. I had tried mothballs and cayenne pepper spray, both suggestions from Ken Armitage, a professor from the University of Kansas and an unparalleled expert on marmots. Neither worked.

Marmots, for those fortunate enough never to have observed one, are large, fossorial, colonial rodents that live at high elevation, having a circumpolar distribution throughout the mountainous parts of North America, Europe, and Asia. They often appear cute, innocent, and playful; in reality, picture the gopher from *Caddyshack*, only three times as big and ten times as vindictive. The creature of my nightmares was *Marmota flaviventris*, the Yellow-bellied Marmot, which has coarse fur that is brown to yellow on the back and yellow to orangish on the belly. Yellow-bellied. The name belies the inner cunning and deceit found within the hearts of all marmots.

Why would marmots attack my truck, anyway? Several friends at the RMBL were convinced it was a direct consequence of my sordid anti-marmot past. Or so they said. It is true that during my formative years I was responsible for the demise of the occasional groundhog from my aunt's garden in West Virginia, for which activity I used a variety of .22-caliber weaponry. It is also true that marmots are functionally alpine groundhogs (or groundhogs are rural, low-elevation marmots—your choice). As much as Ken and other marmot researchers would like to deny it (an easy way to get into a fight with a marmoteer), groundhogs and marmots are virtually identical in their DNA, if not their pelage. So one could surmise that the marmots were taking it upon themselves to get revenge for my years of persecuting their cousins from the East. My nonhunting compatriots reveled in my situation and frequently taunted me with questions about how the marmots were doing up valley, whether my truck would run that day, and what the score was (you know: Marmots 2, Whiteman 0). The copious dust on the Toyota's cab window was frequently fingered into "MARMOTS RULE" and "MAR-MOT MACHINE" and other such niceties. No one else at the lab was having this problem (or would admit they were); it must have been the scarlet letter of my past.

I am rarely superstitious about most things (except football, hunting, fishing, and, well, writing), but the marmots began to have me spooked. My car was a mess this time, and I decided to spend some of my precious research money to get the engine rewired. This necessitated that I temporarily borrow someone else's car. Scott Wissinger, my mentor at the time and, remarkably, still one of my closest colleagues, was gullible enough to lend me his Land Cruiser, thinking that my bad karma would not transfer to his vehicle. He was convinced that it was my truck the marmots were attacking, not me per se. It was easy to agree, as I needed his truck.

On our next trip I openly prayed that the marmots had taken the last few days off. As Andy Bohonak and I approached the Land Cruiser after three days at the ponds, two young marmots fell out of the engine block and ran down the road. "YOU BASTARDS!" I yelled, chasing the marmots and throwing as many stones as my tired arms could muster. "AAUGH!!" Yet the adventure had just begun. After returning to the truck, I popped the hood and there, sitting on the air filter casing in all its putrid furry regalia, was a rather startled-looking marmot. "YOU MOTHER FUDGER!" I screamed, except that, as in *A Christmas Story*, I didn't say *fudge*. Such cursing had become a common part of my vocabulary, and also an easy way of venting my frustration without death and destruction. The marmot quickly dove under the engine but did not appear on the ground to run away. It was hiding. Under the engine.

I jumped on the ground and scooted on my back to see underneath the truck. The marmot was there, sitting on top of the transmission. I grabbed a ski pole we use for hiking over snow and prodded the marmot. It immediately moved into a crevice between the transmission and the engine. I prodded some more. It moved deeper. Certainly, the aggressive poking strategy was not going to work. I chose to back off, hoping the marmot would pop out and escape on its own. We waited a few minutes until my frustration with marmots had stewed long enough. I had heard stories from Ken about marmots that had hitched rides in vehicles for hundreds of miles. This marmot could sit tight for hours, days, months!

I grabbed a pair of leather gloves from the back of the Land Cruiser and rolled back under the truck. I found the marmot's tail and with one sudden pull tossed it out on the ground. It immediately jumped back into its hiding place within the truck. "You have GOT to be kidding me," I muttered. I tried again—pull, marmot out, marmot back in. Clearly, teamwork was needed here. I pulled, Andy poked with the ski pole once the marmot was clear, and finally the beast was on its way back to its burrow. BASTARD.

Luckily, Scott's truck was relatively unscathed and started right away. Apparently, there is a reason why Land Cruisers cost more than regular Toyota trucks. Who wouldn't pay extra for a marmot-resistant engine? We made our way home, and I confessed to Scott. After checking the engine carefully, we were both convinced that little if any harm had been done. Nonetheless, Scott was not interested in having me borrow his vehicle during the rest of that summer. Go figure.

I knew that my new wiring would be in splices if I allowed the marmots another chance. A drastic solution was called for. With some wire cutters, a pair of pliers, and some hardware cloth, I fashioned a marmot-proof fence. During my next trip we carefully set up the fence around the truck, completely surrounding it with three-foot-high hardware cloth and placing rocks in strategic places to keep it standing and to block any holes. It looked strange, we got lots of stares from passing hikers, I was the laughingstock of the lab, and it took some time to put up and take down, but it worked. The marmots never penetrated the fence, and the time spent arranging it was more than offset by a consistently working vehicle.

As it often does, time unveiled the real reason for the mysterious marmot attacks. My graduate work had been done at Purdue, and I made the yearly migration from Indiana to Colorado in my Toyota. Of course, as a graduate student I never had the time or gumption to wash my car, particularly the undercarriage. Heck, the dirt was the only thing holding the truck together, an observation that is still true today. It was not until I moved to South Carolina for my postdoctoral work, and made my first trip from my new home in the South to the RMBL, that I

suddenly was no longer a target of the marmots. I discovered this reve-
lation only after forgetting, in my haste to get to the ponds, to put up the
marmot fence (how could I forget?). Apparently, it was not groundhog
revenge at all but rather the salt on my car that had accumulated, even
on the engine, from years of driving throughout the Midwest. Because
South Carolina roads rarely attract snow, and thus are never salted, rain
and water on the road had eventually cleaned my truck of salt, and I was
cured. Scott's truck was likely victimized for similar reasons, although
he has always been more proactive about keeping his vehicle clean,
which might explain the trivial damage he experienced. A discussion
with Ken revealed that he had suspected the salt phenomenon all along,
but for some reason, it never came up in conversation. Hmmm. Perhaps
he spends too much time with marmots. No matter, as I was now convinced that the curse was over.

Once I was clean of the scourge, I reveled in watching marmots
climb out from underneath Subarus from Ohio and Chevys from Wisconsin. Over the years, I saw scores of marmots running from such
vehicles. Of course, many of these cars were in place for only a few
hours, rather than days, and so they were likely never affected to the
same level as my Toyota. Nonetheless, I chuckled over the possibility
that the marmots were secretly creating havoc in these unknowing
engines.

My gloating was short-lived. The marmots had retreated from my
truck, but they continued to attack us in other ways. They were certain
to continue the war on other fronts, as they had for years. I have thus far
failed to mention that while I was dealing with my numerous car problems, the marmots were hitting me at our research site as well. Different marmots, same nasty habits.

An old mining cabin had been renovated in the 1970s to house
researchers and protect equipment at our study site. We used the cabin
extensively but spent considerable time sweeping marmot dung. Each
year we quickly plugged the largest holes, but young marmots always
found a way into the cabin and targeted anything we had left unpro-
tected. For oversized ground squirrels, they certainly could climb well.
A set of shelves lined half the cabin's west wall, and the marmots
appeared to enjoy running down each shelf and knocking glass vials,
thermometers, plastic containers, and other supplies onto the cabin

floor. For weeks, we would pick up the supplies, put them back on the shelf, attempt to fix the entry holes in the cabin, and the next day find more carnage. The marmots also managed to bite holes in our rubber boots, tear our clothes, eat our data sheets, and generally defecate and urinate on any surface of the cabin that we could use for research. It always seemed that my possessions were most affected, although paranoia could have clouded my evaluation of the data. Only recently have we had the funds to renovate the cabin so that the marmots are completely excluded. For now.

Of course, any belongings left outside the cabin, even briefly, were fair game for the marmots. One of my former assistants, John Gutrich, remembers fondly how the marmots ate the sleeve off one of his shirts while he was working at the ponds. I have had the same experience. Luckily, we were not wearing the shirts while the marmots were chewing up their cotton supplements. John was also present when a marmot ran out in front of us while we were measuring salamanders. Such measurements are a delicate process, and once I get started on an animal I don't like to interrupt the routine until it is safely back in a holding bucket. The marmot seemed to sense this (perhaps the pet psychic could help here) and went right for my inflatable raft, chewing a hole in it and fleeing for its rocky burrow before I realized fully what had just happened. As the raft slowly deflated, I muttered a few choice expletives. If marmots don't count as a plague, then what does?

Thank God for predators. I have quickly realized that foxes, coyotes, eagles, hawks, and especially badgers are my best friends. Badgers are marmot specialists, typically moving up valley and cleaning out one marmot colony after another, often usurping control of the underground burrows after eating all the local marmots, and using the burrow for their own reproductive efforts. When a badger appears near the lab, which is sadly infrequent, the marmoteers are nervous and I am rejoicing. One less marmot, one fed badger, one happy Whiteman.

Another honorable yet infrequent predator in the area is the fox. One year a pair of foxes took over an old marmot burrow right in the middle of the RMBL. They were driving the ground squirrel researchers nuts because once their four kits appeared, they were consuming significant numbers of marked study animals on lab property. One day, soon after young marmots started appearing outside their burrows, about half the lab's human inhabitants were waiting on the deck of the dining hall for dinner to be served. Suddenly, a cacophony of marmot alarm calls went up, basically a series of short, high-pitched whistles that warn other marmots of approaching predators. Apparently, the alarm calls were a bit too late, for as we watched, one of the foxes emerged from underneath a nearby cabin with a young marmot in its jaws. It trotted rather triumphantly down the trail, past the dining hall, and across the county road to its burrow. While a group of three marmoteers followed in hot pursuit (just wanting to see the fate of the marmot, and perhaps its identification marks, I am sure), three-quarters of the assembled crowd let off a large cheer. Some new students, with less knowledge about the ecological importance of predators and how rare it is to see a predation event, and certainly no understanding of the evil ways of marmots, were saddened by the young marmot's fate. Neophytes. The rest of us celebrated, and the foxes were a special part of the lab for the rest of the summer. Heck, we made T-shirts with foxes on them that year.

The most amazing thing about human-marmot interactions (think of the parallel human-Ebola interactions) is that some of my colleagues at the RMBL actually study marmots. For a living. They somehow enjoy watching marmots through 40x spotting scopes for hours on end and will revel in what marmot 459 did at the local colony. "Boy, was that exciting," they will spout without prodding. You would think they would realize that they are functionally studying a four-legged, furred virus, but apparently not.

The king of these obviously demented souls is Ken Armitage, who has been studying marmots in the area for more than forty years. And yet through all this time he has somehow grown fonder of marmots rather than realizing the incredible mistake he made during his first field season of watching them. Many of Ken's students managed to get an M.S. or even a Ph.D. while studying marmots. Earning a Ph.D. in field biology is bad enough, but attempting to conduct your dissertation research on marmots is like sticking your bare arm in a nest full of fire ants and expecting to come out unscathed. And yet many of Ken's students not only survived their graduate research on marmots but flourished afterward, becoming some of the finest ecologists and evolutionary biologists in the world. Interestingly, though, few still study marmots. Some still study small mammals, but others have diversified into fish or even plant-insect interactions. It could be that Ken's students simply broadened their horizons after their graduate work. An alternative hypothesis is that the marmots drove them to these other, more viable, safe, and, let's face it, reasonable fields of study.

Some marmoteers, however, have conducted absolutely bizarre research on the animals, making me think that the marmots actually have limited powers of mind control over some humans. Dan Blumstein, now at UCLA, conducted his postdoctoral work with Ken by observing the messages (via alarm calls) that marmots were sending to each other in response to artificial predators such as toy gliders shaped like eagles and the pièce de résistance, Robobadger. As you might guess, Robobadger was a remote controlled car engulfed by a stuffed badger. Very realistic, to be sure, but what was Dan thinking? Sure, he got some nice publications out of this study, and landed a great job, but did he ever just sit down and think, I am playing with toy cars to study MAR-MOTS? Perhaps the marmots were simply ordering Dan through subliminal or telepathic messages to give them some entertainment. Give UCLA credit, they did not hold this obvious mental disorder against Dan when he was interviewing for the job.

Brett Woods is another interesting marmoteer. He studied the feeding preferences of marmots for his dissertation work with Ken. I could have saved him a lot of time—they like 1985 Toyotas from northeastern states, mostly the wires but an occasional hose now and then to keep them regular. They enjoy supplements of cotton shirts, rubber boots and rafts, and data sheets. No study needed. He never even approached me but rather wasted years of his life figuring out the optimal diet selection of these fat furry feeding freaks. For science. Of course, the marmots likely had control of Brett just as they manipulated Dan. How else can one explain his daily behavior of carrying pounds of horse chow and dandelions to marmot colonies every day during the summer months of the year? Sure, he landed a great postdoc at Northwestern and then an awesome job at Beloit College, but it cannot take away the gruesome truth that he studies *marmots*, and that he will carry those scars with him for the rest of his life.

Like most curses, one has to learn to live with marmots. Besides living in the mid-South, washing my truck, and making sure the cabin was marmot-proofed, I have done several more extravagant things to maintain a truce with marmots. For one, I married an academic grand-daughter of Ken Armitage, her Ph.D. adviser having conducted his own dissertation research on the same marmot populations whose descen-dants had tormented me all those years. Jerry was smart and got out while he could, but nonetheless I believe his *chi* has been passed on to me, as sort of an existential dowry. My wife, now an associate professor of biology, teaches mammalogy, which has to alleviate some of the angst the marmots feel toward me since she espouses the supposed impor-tance of small fossorial mammals to her students. And, perhaps most im-portant, I gave up on groundhogs long ago. It probably doesn't hurt that my Tennessee home is on the edge of the woodchuck's range, and I have seen exactly one groundhog within a three-mile radius in the past seven years. I couldn't have picked a nicer place to live, but I still make an effort to celebrate Groundhog Day. The marmots might be watching.

6

In Big Fence Country

MICHAEL ROGNER

Two days after we find the tracks, Rob, the black-sheep brother, stops by and asks if we've seen anything.

"Not really," I say.

"No wolves?"

I shake my head. The prints were dried up on the little two-track that cut across our plot. They were as big as my hand. The Bureau of Land Management has closed the road, but we have a permit, and all the ranchers know this. Rob points to Dilworth Bench, a large, flattish, sage-draped area on the flank of the mountains to our west, to show me for the tenth time where the pack has been summering. Our study spreads below it. Later on I'll discover that he's already been out to our plot searching for tracks, so he knows I'm lying when I insist that we haven't seen anything exciting.

"Some of our birds got eaten by rattlesnakes," I tell him.

"What do you do?"

"What do you mean?"

"You shoot them?"

"Shoot what?"

"The rattlesnakes."

We don't. For my part, at least, I kind of like them. It's tough not to like anything strong enough to live in the arid sagebrush scrub of south-central Montana. They're Prairie Rattlesnakes, a subspecies of the Western, and when you step too close to one that rattles, it reminds you in the simplest terms that you're human.

"Yeah, Dad don't let us shoot them either," he says. Then we look off toward the mountains and I wait for him to ask me again about the wolves.

Rob is the only son who's left the ranch. He does campground construction around the "Park," meaning Yellowstone. The house we're staying in is a guest place on his parents' ranch. The Clark's Fork Yellowstone flows past our back door. I've tried to stand on the porch and throw a stone into the river left-handed but can only make the irrigation ditch a few feet short.

"I'm surprised you haven't seen anything," he says.

I shrug.

"No griz or nothing, huh?"

"We're always looking down," I say, pantomiming nest searching among sage. "Probably wouldn't even notice."

Rob comes around only a couple times a month. He'll return home for a weekend and I'll see him out on a tractor, and then he'll mill around in front of our place until I invite him in. Or I'll go out and offer him a beer, which he'll decline. He's in his early twenties, small and wiry with a sparse goatee, and he's always wearing a grimy baseball cap that he removes when he enters the house.

The problem, he says, is that the wolves are getting too smart. They know we can't shoot them, so they've come out of the Park and started eating cattle. Every time we talk he's got a new list of figures about so-and-so losing 46 head and his neighbor losing 25. Of course, this is all bullshit, a point my boss urges me to bring up, but if you let him go on long enough—and this takes patience—he'll come around to areas where we share common ground. The problem, the real danger to ranching and the reason he left his family's ranch, isn't just the government regulations that say, for example, that shooting endangered animals is illegal; it's more the collusion between government and the small minority of super-wealthy landowners.

"You got to remember," he once told me, "that this country was founded on a revolution. Not a revolution of folks like you and me, but a revolution of rich fucking landowners who didn't want to pay their taxes."

Rob tells me about their place down in Wyoming that they've been trying for years to "square off." They own seven and a half sections of land (a section is 160 acres) and have been trying to purchase the last half-section so that the entire ranch could be one series of long, straight lines, which would make planting, fence building, and everything else easier. But they're stuck on a list and will be for decades.

"But this guy," he says, pointing east to a line of tan, rocky hills, "he wanted to square up some land, so he made a phone call and everything got done. He's got connections. He knows Conrad."

Rob spits by our feet then quickly kicks some sand over it. "Don't make it right. Fuck." Then we watch the bench again, and sooner or later Rob starts talking about the renegade wolves.

"Conrad" is Conrad Burns. He's the junior senator from Montana. The land Rob referred to belongs to Earl Holding, owner of Sinclair Oil. Holding was once the largest landowner in the state, before the days of Ted Turner. He also owns huge tracts in Wyoming and Utah. Salt Lake–area environmentalists refer to Holding as "Satan, but richer."

The oil and gas industry gave $160,000 dollars to Conrad Burns's last campaign.

I had never been to Montana before I worked there. I was prepared for the endless rivers (the Yellowstone is the largest undammed river in the lower 48), the views of mountains, and antelope and mule deer wandering roadside, but I wasn't prepared for the fences. Big Sky country is really Big Fence country.

Rob told me about a local rodeo star who dropped out after too many injuries and moved on to bigger and better things.

"He could see what was what," he told me.

"What do you mean?"

"People, rich folks, Hollywood types and whatnot were all looking for land. Mel Gibson, the movie star, he's got a place not far from here. And what do they all want when they get here? Fences. Keep out the riff-raff. So Dane stopped the rodeo and started a fence-building business. Doing good too."

Mostly the fences are the same. Horizon to horizon, steel rods with four strings of barbed wire running taut between them climb hills and drop into dry washes. On government fence the bottom strand isn't barbed, but it is on private. There's a law, one of the many unenforceable laws that govern the meeting places of public and private land, which states that the bottom strand can't have barbs. Pronghorn don't jump fences like deer. They'll cruise along at 40 miles per hour next to the fence and then in one quick motion drop down and shoot beneath it. Pronghorn numbers are up in Montana, but not high enough to be letting barbed wire rip apart their backs.

"Of course, anything else can get under that fence," Rob says. "Calves, sheep, coyotes. Wolves."

Everywhere, fences fragment the land. And numerous times, out in the sage a mile or more from the nearest road, I found rolled bales of wire taller than me, waiting.

I was there to study birds. Brewer's and Vesper Sparrows, Lark Buntings, Sage Thrashers, Loggerhead Shrikes, and Lark Sparrows. Lark Sparrows are largish cream-colored sparrows with a dramatic chestnut facemask and a single black mark on the breast. Their voice sounds like R2D2, the robot from *Star Wars*. It's impossible to tell the male and female apart unless they're in your hand. They nest mostly on the ground, and one day my co-workers found three of their nests hidden beneath Wyoming sage (these were the first three we'd found). My boss said I'd have no trouble finding some on my plot, which is when I knew I was in trouble. The sparrows would hop around openly with grass in their beaks, and with a bit of patience, she said, they'd lead me right to the nest they were building. If I missed that, they were just as easy to spot when feeding nestlings.

I found my first Lark Sparrow nest by accident. I was checking on a Vesper Sparrow nest, another ground nester, which I'd located once and couldn't seem to find again. We'd devised all sorts of tricks with colored flagging and GPS units, but still, nests were always hard to find the second time. Another problem on that particular plot was the occasional herd of cattle that would pass through. Since they're dumb enough to eat anything, flagging was naturally on their menu. I narrowed the nest area down to about one hundred square meters, and so I crawled from shrub to shrub. After an hour I found a perfect nest cup hidden beneath an old, dying sage. This jogged my memory of the nest

I was looking for—the "look" of the shrub it had been beneath—and this definitely wasn't it. So I flagged this one. Two days later an egg was in this nest, large and splotched with purple—a Lark Sparrow egg. I also found the remains of the vesper nest five meters away. It looked like a coyote had gotten it.

My second lark nest was just as much an accident. Walking up the wash to check on some Brewer's Sparrow nests, I heard a bird flush from right under my feet, usually a sure sign of nesting. I poked around in a lush silver sage and found the nest beneath it with five eggs.

My boss was happy, but I wasn't. There was a lot of Lark Sparrow activity farther up the wash, and I couldn't find a single nest up there. I'd watched one pair covering the top of an old magpie nest with grass and thought I was on to something special, but in a few days all the grass was dried and the birds no longer showed interest. Then one day I finished my work by ten-thirty in the morning (we got into the field at five) and decided I'd observe those larks until I saw the white truck coming to pick me up. It was hot out, ninety degrees or so. I walked within twenty meters of the wash and crawled to the edge. I had a perfect view and excellent binoculars. I took off my shirt, reclined against my backpack, and waited for something to happen. Two minutes later a Lark Sparrow flew in and perched atop a dead juniper. It held a grub in its beak. This was too easy.

Then a second bird flew in. They perched a foot apart and stared at each other, tilting their heads side to side. I watched. They waited. The sun was getting higher, which meant it was getting hotter, but I wasn't going to lower my binoculars for a moment to wipe the sweat off my face or look for the truck. The birds waited. Ten minutes passed. I knew they were patient birds from past observation and from trying to catch one. We'd set up a mist net at one of the other nests and a parent came zooming in with food for the chicks. It stopped short of the net, perched atop a sage, and waited for forty minutes until we went away.

"Just show me your nest and I'll go away," I whispered.

The sparrows waited.

"Aren't your nestlings getting hungry? That sure is a nice, juicy grub you got there."

Waited.

I talked, and they waited, and then I heard two more Lark Sparrows on the ground beneath the tree. I held my binos far enough from my eyes that I could still see the two in the tree but could also look quickly to pick up the other two. They were feeding. I decided to ignore them and focus on the first two. I had all day. Sure it was hot, but I was patient, and I had the desire and the perfect location. It was a matter of time.

Then this happened: the two in the tree flew up a few inches and came together at the breast. They spun in rapid circles and then landed again. I wasn't sure what exactly had transpired. Either they had switched places or there had been a mid-air food swap. I sat up straighter. My arms were starting to ache from holding the binos so long, but I wasn't giving in for moment.

"Nice trick," I said. "I'll give you that one, but I'm still going to find your nest. Why don't you just show it to me, and we can all move along?"

Five minutes. Ten minutes. I took turns holding the binos one-handed, resting each arm. Fifteen minutes. Should I watch the other two? Were they still there? Lark Sparrows are polygamous, so it's possible there were three nests in this area. I kept on the original two. Twenty minutes.

Pay dirt. The one with the food flew to the edge of the wash and hopped beneath a rabbitbrush. Then hopped right back out again. I almost ran over.

"Patience, Rogner. Keep observing."

Sure enough, the sparrow still had the grub in it beak. It hadn't fed anyone. It went into a sage, briefly raising my hopes again, but it dashed them by climbing limb to limb until it popped out on top. And it waited. It waited ten minutes. Then twenty. I wanted to throw rocks at it. Then it flew back to the juniper, next to the other bird, which now had food in its beak too. Somehow I'd missed that. But these small surprises only reinvigorated me. The nest was there. I just had to watch. They played their waiting game, but I was oblivious to it by now. I could have waited all day. I could have passed the night there. It was me against these two birds, and I was certain I was the smartest of the group.

When they flew, they went in opposite directions, but I was prepared for that. There's a trick to watching one bird through the lenses and another with your naked eye, and I was good at it. I was prepared. The one that went right circled over the wash and landed on the ground next to the one that went left. They perched on a fallen branch, and I knew the nest was nearby. They waited, and my shoulders were getting sunburned, but it didn't matter.

Then I heard something. I'd been worried about rattlesnakes crawling under me for shade, but what I heard was ominous in a completely different way. Our plots were on the east slope of the Beartooths, a mountain range that doesn't look inviting in any sense. It's a wall. Come in if you dare. What I heard sounded like a steam locomotive. It was the wind, I knew, rolling down off those mountains behind me. And a lot of it. I could hear the tortured strain of shrubs fighting it. I turned just as it whipped over the wash and slammed into me. With it the temperature dropped twenty degrees, and there was a wall of thick, charcoal-gray clouds storming along above it. All this—the turning, the looking—took five, maybe seven, seconds. The white truck was just pulling in, silhouetted by the clouds. The roads we drove to get here were solid clay, and it took only a sprinkle to make them undriveable. We'd learned our lesson a few weeks before when the truck did numerous 180s but somehow managed not to get stuck.

Anyway, I turned back to my birds. Five, maybe seven, seconds out of an hour and a half of observation, and they were gone. What else could I do? I could either smash my binoculars on a rock or laugh, and my binos are a shade too expensive. As the rain started pelting me, I searched the shrubs the birds had been near but found nothing. So I packed up and trudged the half-mile in blistering wind and rain.

There were five of us in the truck, and we all swapped stories, and my boss, encouraging to the end, said that I'd find the nests. She added that all the behavior I was witnessing was good for other parts of the project. Sure, it was all crap, but I was cold, wet, and sunburned, and I took whatever praise was offered.

The youngest ranch brother, Mike, was in charge of the sheep operation. He wasn't the politick that Rob was. He didn't have his sensitivity. A baby black bear, he told us, once wandered past their kitchen window while they were eating dinner.

"What happened to it?" I asked.

"We made a rug out of it."

Rob told me the same story. "I just don't get it," he said. "But fuck, I ain't here no more. Dad won't let us kill a snake." Then he turned away and kicked at the dirt before he said anything bad about family.

Mike also told me about letting his little sister run the hay bailer. "She ran over seven or eight fawns, and she sure don't want to drive it any longer." Then he laughed at his sister's bleeding heart.

Mike was interested in what we did, though. He came out into the field with us twice, and these were chances, according to my boss, to make a real difference by showing him another perspective, and I agree. Knowledge changes things. It lights the way toward hope. She even let Mike handle some birds.

To return the favor, he taught us how to shear sheep. It's one of those weird highlights of a summer in the field—five urban biologists in the slanting sunlight beside a river fighting with 150-pound sheep. The cameras were flashing.

We also got to take Jay Parks, the BLM biologist, into the field for a day. He talked about prescribed burning of the Big Basin sage to increase foraging for cattle, and my boss had the ready answer. She led him deep into the thickest sage and showed him a Brewer's Sparrow nest with four eggs.

"I didn't even know birds used this stuff," he said.

He also didn't know what kind of birds were out there. Being the wildlife biologist on BLM land in Montana isn't something they'll make a TV show out of. He has way more land than he can handle, and it's all about cattle and how to maximize the land's ability to handle cows. Cows are the good "wildlife." Everything else is bad or, at best, neutral. And Jay gets it from both sides. The ranchers want unlimited, virtually free access to public grazing lands, a privilege they now enjoy, almost. Others want a more balanced approach toward the interest of ranchers and the interest of healthy public land. And serious environmentalists want the cows and miners out altogether, the fences gone, and the wildlife back. It's public land, they argue; it shouldn't be turned over to a few wealthy ranchers and mining corporations, especially without any enforcement of regulations on how they treat it. As it is, BLM land in Montana is poisoned and festering, and it's no surprise that the populations of the birds we're studying have been plummeting for as long as anyone's been paying attention.

A conservative guess is that there has been a 2 percent decline in birds every year for the past thirty years for that particular habitat. That doesn't leave a whole lot of birds or much hope for their future. Overgrazing. Urban sprawl. Pesticides. Lack of water. You name it, and they're facing it.

But some birds are doing well. You look up into the big sky of Big Fence country, and you see birds. Feedlots are Gardens of Eden for Brown-headed Cowbirds, pigeons, and blackbirds. As the biodiversity decreases, their numbers increase. Look at the pigeons in any major city. Our wildest, most remote places aren't much different. We're losing our birds, and they can't compete for the limited resources Americans already put toward the recovery of large mammals. Sparrows aren't nearly cuddly or awe-inspiring enough to make for a heartrending commercial.

Rob says most of the problems can be traced back to the Park. It's too crowded with tourists, he says. We're standing by the cattle guard in front of the house.

"How do you fix that?" I ask.

"You don't," he says. "They aren't interested in fixing it. Every car that goes through that gate is twenty bucks in somebody's pocket. Plus you've got food and hotels and knick-knacks, and you can figure someone is getting rich. A lot of someones. It's tough on the wildlife, though. No wonder they want to leave."

"What about problem wildlife? The ones that bother the visitors?"

He points to the southeast, to a line of mountains rising just over the horizon. "They cart them off to the Pryors. It's about the worst place in the world you could put them, but they do it because nobody has any money over there to oppose them. Just a bunch of Indians. And the places they should take them are full of movie stars and other rich people who don't want wolves and bears in their backyards. So what happens?" He pauses to spit. "Shit, they don't stay in the Pryors. They know where they've got it easy, so they head back to the Park and come right across these ranches. The Feds say we can't do anything about it, but they won't even put them in the right place."

"What about the Park? The overcrowding?"

"What I'd do, not that they'd ask, is start a tram system. Then they'd get the same number of visitors, and all the right people would still get rich, but there wouldn't be all those cars. How much pollution you think comes from those cars? That place is one big traffic jam. It's like a city. And they won't let me ride my snowmobile in the winter because they say it pollutes things up. Shit."

I ask about the snowmobiles, and his answer takes us back to wolves. I tune out for the next twenty minutes and look for an opening where I can steer him back to something else.

7

No Forwarding Address for Vernal Pools

ROY A. WOODWARD

In the early 1990s I was conducting surveys for vernal pools in the foothills of central California. Vernal pools are seasonal wetlands that contain an amazing array of colorful spring-blooming flowers and interesting invertebrates, many of which are listed as threatened or endangered. The loss of vernal pool wetlands in California has been astonishing, with over 90 percent of these important habitats destroyed, mostly by urbanization and plowing.

Armed with a set of topo maps, I would drive as near as I could get to likely-looking grasslands and then crisscross the land on foot looking for the shallow depressions and whatever species the pools contained. Most of the land in this area is divided into large, private cattle ranches surrounded by barbed-wire fences. On occasion, when no one was around, I might hop over a fence to conduct a quick survey, and no one would know the difference. However, if there was a ranch house nearby, I would go ask permission to look around, and landowners were generally very cooperative and interested in what I might find on their land.

On one occasion I visited a ranch where the house sat on a small hill that overlooked the surrounding area, so I had no choice but to go ask permission to conduct my survey. A large, good-natured older cowboy came to the door, and we stood on his porch as I introduced myself. He cordially asked what I was doing, and I said, "I'm looking for vernal pools."

The cowboy cocked his head to the side, tilted his Stetson back on his head, and stared out into space for a few moments. Then he replied, "You know, I don't think old Vern lives around here anymore."

After a bit more explaining, he willingly granted me permission to conduct my survey, but I was struck by the unfortunate nature of his statement. I realized that, within a few years, development would surely continue to consume seasonal wetlands, and "old Vern" probably would be gone for good.

8

Running the Wind

JENNIFER BOVÉ

The Wind River is cold. Not the kind of cold that begs a swim on a scorching summer afternoon (there are relatively few of those in the Pacific Northwest), but the kind of cold that seeps down the neck of your wetsuit and freezes the very fluid in your spine. The waters of the Wind course from high, quiet places in the Gifford Pinchot National Forest where the snowpack doesn't melt until well into June. Crystals of ice succumb reluctantly to thaw and dribble through cushions of moss into rivulets that quench creek beds and thirsty mainstem tributaries like Paradise, Dry, Trapper, Panther, and Trout Creeks. This aggregation of icy water brings with it tiny particulate hints of summer's emergence in the high country and welcomes spawning summer Steelhead into the Wind. Even on a bright morning in August the water sank needle-sharp teeth into my skin, but my anticipation didn't falter. I couldn't wait to slip into the wide, turbulent pool at the base of Shipherd Falls and start snorkel-surveying for Steelhead.

I looked over at two of my Forest Service teammates, who were kneeling at the water's edge. Becca smiled a silly frog-faced grin at me beneath the snug rim of her dive mask, and Nate stuck his thumb in the air. I knew they were itching to get started as badly as I was. Landy, however, was still maneuvering over the wet rocks on the bank, clutching a snorkel in one paw and balancing with the other.

"Hey, Landy," I yelled. "What's the matter? Afraid you're gonna fall and end up with another crack in your ass?"

He smirked and gave me the finger.

Landy was usually the self-proclaimed balls and brawn of the crew, so in those moments when we caught his bravado going a little limp, he knew he was fair game for ridicule.

Once he clambered around the last rock ledge to join us at the pool's edge, I skimmed over our course of action, shouting above the roar of the falls. We were to line up across the river and count every adult Steelhead in our paths, noting any ID tags or missing adipose fins.

A clipped adipose fin is an identifying mark used to differentiate hatchery Steelhead from those that are wild. The Washington Department of Fish and Wildlife stocked the Wind River with hatchery-bred smolts for more than three decades until, in 1998, the newly recognized threat of hybridization halted the operation. This kind of anadromous stocking program has been the government's solution to the unpredictability of runs and the decline of salmonid populations since the construction of the first Pacific Northwest hatchery in the late 1800s. Captive propagation has been misrepresented as an almost natural panacea for far-reaching injuries inflicted on watersheds by a hundred and fifty years of unbridled industrialization.

In reality, however, the hatchery programs are another strike against the survival of wild Steelhead. The threats posed by a hatchery Steelhead cannot be identified by such environmental red flags as concrete or erosion, but the fish's genetic inferiority, resulting from captive rearing, is every bit as dangerous to the wild population. Hatchery Steelhead carry communicable diseases that inevitably run rampant in feeble captive stocks; hatchery Steelhead mate with wild fish, diluting what's left of the more diverse and vigorous native genes; and hatchery Steelhead now outnumber wild Steelhead in the Wind River. Under their guise of likeness to wild fish, the hatchery brood are quietly displacing native Steelhead and simultaneously replacing the river's natural heritage with an inferior run. Since genetic diversity is the shield behind which individuals can persevere in the face of environmental change, the tragic loss of diversity among the Wind River Steelhead threatens the long-term viability of the population as a whole. In recent years only about fifty pairs of adult spawning summer Steelhead that descend from the ancestral run have returned to the Wind, a mere handful of a population that historically distinguished this river as one of the most plentiful Steelhead fisheries in Washington state.

As part of a collaborative effort on that mild summer day, my crew and I joined biologists from the Washington Department of Fish and Wildlife and a handful of other partner agencies to count all the Wind River Steelhead—wild and hatchery stock alike—in order to obtain an accurate census of the river's population. I adjusted my mask and wriggled the mouthpiece of my snorkel between my teeth. It was one of the few functional snorkels that Becca and I had found buried in a bag of old, moldy wet suits at the ranger station. The Forest Service isn't keen on buying new gear for summer crews, so we'd picked out the least offensive articles that were anywhere near our sizes, patched the obvious holes, and did our best to scrub them clean.

"Let's stop after every pool to tally the fish, and I'll record them," Nate yelled. He was wearing a plastic wrist plate and grease pencil for that purpose.

We nodded to one another, and I plunged in first. I attempted a vicious crawl stroke to beat the current but swam more like a marshmallow in a snowsuit. I managed to power across the pool through the glitter of tiny bubbles that boiled in the rush of the waterfall, but I started to drift fast and couldn't stop to check for the others. It was impossible to fight the roiling surge from the falls to look for fish; I just needed to make it across.

Within ten feet of the far canyon wall, I stopped struggling against the current and turned to face downstream, giving the team a quick wave, and then I stretched my arms straight out in front of me and lay my face into the icy water, suddenly alone in the cavernous beauty of the pool. It was quiet and ethereal, and I floated effortlessly in the buoyant neoprene wet suit as if gravity had set me free. Soaring some thirty feet above the floor of the clear blue pool, I remember thinking, *This must be what heaven is like*. I was an angel on the wing.

I peered intently into the pool's depths. Down, down my eyes slid along the massive rock wall, searching its dark crevices for a sign of the elusive *Onchorhynchus mykiss*. Steelhead are known to snuggle into strange places at cockeyed angles; you have to look carefully. Old metal grates were strewn haphazardly among the rocks at the bottom of the pool, but they harbored no fish that I could see. Once part of the Shipherd Falls fish ladder, the grates must have been dropped and were too much trouble to retrieve.

The fish ladder alongside the falls was constructed to permit the passage of hatchery-bred spring Chinook Salmon past the steep series of cascades so they could return to the Carson National Fish Hatchery, their natal facility, eighteen miles upriver. Before the ladder was built, these falls were a natural barrier to the heavy-bodied Chinook. The only fish that had ever been able to scale the falls was the sleek and stream-lined Steelhead, and so it is the only native anadromous fish of the upper Wind.

I was so intent on studying the spaces within those metal grates that I barely caught a glimpse of a shimmering silver torpedo darting from beneath a ledge in the rock wall. I was sure it was a Steelhead, but it moved too quickly for me to check the adipose fin. I wanted to wait for another look at it, but I had to keep up with my teammates to maintain the integrity of our survey, and Becca was already skating ahead of me with the agile speed of a water strider.

At the tail-out of the pool the bottom rose up to meet me as I came in for a landing on the rocks, and I emerged as awkwardly as a beaching seal onto a slab of sun-baked basalt. The bellowing falls swallowed my words when I shouted at the others to join me, so I waved my arms until they took notice, and we gathered on the bank above the first rapid to give Nate our data.

Landy said he'd seen one hatchery Steelhead and a bruiser Chinook. Nate marked a tally on his wrist plate for the Steelhead. Neither he nor Becca had seen anything. I told Nate about my unknown Steelhead, and we all agreed that it didn't look like a very promising start to the survey. We knew we should see more fish than the teams working the upstream reaches. The lowest stretch, from Shipherd Falls to the mouth of the Wind, was where most fish would be lingering on their journey upriver.

We skirted the first rapid because it was short and rough. Even if by some fluke there had been a Steelhead in its torrent, we'd never have seen it. So we took to the water again in the following pool, which was narrower and shallower than the big pool and renowned among local anglers as a spring Chinook honey hole, and we found it littered with lost lures and lead weights. I pointed out four large-scale suckers heading toward Becca's lane, and she gave me a thumbs-up to tell me she saw them.

The four of us had to crowd into a narrow eddy at the pool tail because there was no easy place to get out of the river in that part of the canyon, and with the current coursing swiftly into the next chute, it was a challenge to stand still without clutching onto one another for balance. Luckily for Nate, whose Japanese build was slight and tempting to the tugging current, none of us had anything to report from the pool but suckers, so he didn't have to loosen his grip on the canyon wall to take data.

"Suckers," Landy laughed. "That sucks, huh?"

Becca shook her head. Sometimes there were no words for Landy.

"Get it?" he said. "*Sucks.*"

"Yeah, Landy," she sighed, "we get it."

They started bickering about how we could maneuver along the wall past the rapid, and I was about to explain that we just needed to *do* it when I began to lose my footing. I ground my feet desperately, but the metal cleats on my wading boots that were crucial for walking on buckskin logs clawed uselessly across the slick bedrock. I glanced over my shoulder into the chute. It wasn't a long one, but it had a good drop to it and a number of boulders to dodge.

"Oh, *shit*," I gasped, frantically groping for Becca's arm. Neither she nor the others noticed, though, as I slipped out of their reach and into the white gullet of the rapid. I sucked in a desperate breath and stared skyward, attempting to shield my head with my arms and waiting for water to engulf my snorkel. But the ride was soft for all its fury and speed, and even though my limbs were swept akimbo, I kept my head above the surface enough to see the white belly of an Osprey flying in time directly above me. The water swept me so quickly through the course of boulders that I hardly even had time for fear, and all I could think about is how I wished I could do it again.

As soon as the river allowed me enough control to stand up, I planted my boots and yelled back to the crew, who all looked pretty tiny at the top of that rapid. They waved anxiously and cupped their hands around their ears to show they couldn't hear me over the water.

I gave them a wide, exaggerated wave to follow my lead, and I was nearly set off balance again. "Come on!" I shouted, wobbling in the waist-deep riffle.

Their arms dropped to their sides. I couldn't see their faces well, but I'm sure their expressions were truly comical. Landy and Nate kept looking at the steep bank for another way past the chute, and Becca shook her head from side to side.

I nodded. "Yeah!"

Their lack of enthusiasm was obvious, but they knew I wouldn't let them live it down if they didn't do it. Becca jammed her snorkel in her mouth first and let go of the wall. As the river pulled her down, she slapped her arms protectively across her chest and stuck her feet out in front of her. Next Nate released his grip on the wall and let the water pull him through. Landy, who had undoubtedly been waiting to see if the other two would drown in the rapid, now had no choice but to upstage us all, and he took the plunge like some kind of extreme sport diver—head first. I cringed, waiting for him to come up bloody, but he slid through the chute in a matter of seconds as if on a water park ride.

We all laughed and smacked each other's gloved hands, and our giddiness got us into a game of who could knock the others off kilter in the current. I tugged Nate to his feet before he could float away. Howling hysterically, Becca shoved Landy, who was studying the flock of nude bathers at the natural hot springs across the river, and he nearly went over like a big Doug fir.

When we finally got serious enough to start surveying again, we coasted downriver in a loose raft on the current. At times less than three feet deep, the river was illuminated with sunlight streaming through the clear water. I watched caddis fly larvae in their meticulously constructed casings scuttling throughout crevices in the bottom gravel, and plum-striped par darted timidly from my path. My outstretched forearms served as bumpers as I maneuvered gently around rotund flesh-colored boulders that looked like giant skinny dippers seated in the stream. Occasionally, though, the current played rough. It yanked me through shoots I would rather have avoided and wrenched my body around rocks that didn't budge when they struck bone. Aside from some under-water grunts and cussing, I took the pain in stride, reminding myself that no day in the field would be complete without at least a few minor injuries, and bruises don't even leave scars.

Within the next half-mile we counted ten Steelhead. Seven of them were hatchery fish (of which three carried fluorescent "spaghetti" tags implanted by state biologists for identification), two were wild, and one was scarred so badly that nobody could tell one way or the other. These numbers were by no means encouraging, and we were all starting to run low on humor. Fatigue was creeping through my muscles and into my mood, Landy had stopped talking except to report his fish, and by the time we were within sight of the bend below the high road to Carson, Nate was so cold we had to clamber out onto the gravel bank so he could warm up.

When he yanked off his hood, his teeth were chattering and his lips were a pale shade of purple. Becca and I peeled off our hoods too. Mine had constricted the vessels in my head to the point that it felt as if no blood flow was reaching my brain. My body was tight with ice water–induced rigor, and the back of my neck was sore from craning to see through my mask. I couldn't help it—I was dreading the last stretch. From there on out, because of the Bonneville Dam, the Wind was no more than an unnaturally wide and slow backwash of the Columbia River referred to as the "Bonneville Pool," and it would take hours to survey unless we were to forcibly swim it.

Landy, to my surprise, seemed to be in the best shape of anybody. He unzipped his wet suit, claiming the air would help him warm up faster. I knew he just wanted to show off his bare chest, though, which he did all the time, and I can't in all honesty say I minded when he did it. Of course, he told Becca and me we should do the same.

Ignoring him, we began to speculate about the kinds of goodies that might await us at the bunkhouse barbecue later that evening. There was talk of deer steaks, potatoes, baked beans, and beer until nobody could bear thoughts of food any longer. If we didn't get moving, we feared we might just starve to death right there on the riverbank.

The long hole at the river bend rewarded us for our suffering. There were nine bright, beautiful wild Steelhead basking together in the pool. Granted, nine seems unworthy of mention compared with the three thousand wild spawners that are estimated to have inhabited the Wind just fifty years ago, but it was more than any of us had ever seen at one time, and it felt like we'd struck gold. We cruised over them slowly, and the Steelhead allowed our presence as if we were no more than logs floating downstream. They waved and undulated slightly in their shimmering blue world, and it seemed to me that they must be relieved to have finally returned from the sea.

Those fish were the survivors of an amazing journey—one that we could only imagine. From their redds somewhere in the upper water-shed of the Wind River, they had hatched and grown and persevered for up to seven years before undergoing the maturation process of smolti-fication and following the urge of instinct down the Wind to the mighty Columbia River, past the colossal obstacle of Bonneville Dam, and onward to the Pacific Ocean. Their bodies had changed physiologically to adapt to the briny marine environment, and they had changed again to return to freshwater.

They must have met strange and dangerous foes on their journey. Perhaps those Steelhead had been hunted by sleek black-and-white giants sliding through dark waters in hungry packs. Or maybe they'd passed vast nets where their brethren hung by their gills like ghosts, sus-pended in time. They may have narrowly escaped sea lions and the well-disguised hooks of patient fishermen. The strongest among them would swim on to conquer Shipherd Falls and fight the unrelenting river to get back to the tributary that spawned them. They would find mates, con-struct redds, and spawn a new generation. And then, unlike other salmonids, the most vigorous of the fish we saw beneath us might begin their epic travels a second time the following spring in a seamless cir-cle of iteroparity.

At sundown we gathered around a bonfire that the wildland fire crew had stoked in front of the bunkhouse, welcoming the heat as it soaked into our chilled and weary bones. We swapped stories with the other fish heads about the reaches everyone had snorkeled, and naturally, each of us exaggerated a bit to impress the hydrology kids who'd been stuck entering data in the office all day.

My husband, Chris, also a fisheries technician, brought me a paper plate piled high with exquisitely charbroiled fare. He sat beside me on my cooler, and I draped my leg over his knee, relishing the smooth dryness of my skin against his. He told me that his team had surveyed a section of the upper river, near Beaver Campground, and they had seen a grim total of three wild fish.

Becca sat on the other side of me with her feet resting on her dog, Seal. The shepherd kept her eyes trained on me since she knew I'd be inclined to slip her a bite of my food. Everybody called Seal the "Bunkhouse Fugitive" because the ranger district would have thrown both the dog and her owner out on their tails if word had gotten out that Becca was keeping her there. Luckily, Seal was good at flying under the radar. I tossed her the last piece of my venison and tilted my head back to gaze at the night sky. Glowing embers from the fire fluttered up to mingle with the stars, and for a moment I was in the rush of the river again. As if from underwater I heard Chris describing the bull elk carcass his team had found in a logjam, and I heard Landy bemoaning the Bonneville Pool.

Awash in my own exhaustion, I suddenly realized how deeply I admired the Steelhead we were there to study. The relentless drive to survive did not let them rest long in their travels from the Wind River to the sea and back again; theirs was a constant struggle for the possible privilege of contributing genes to the next go-round. Maybe our lives as seasonal techs were somehow analogous to those of the Steelhead. We bobbed in eddies of temporary jobs, circling from place to place, swimming onward for the sake of motion itself. Call me crazy, or just too damn tired to make good sense, but there was some sort of justice in the parallel.

I focused on one red ember and followed its blazing course up through the river of campfire smoke until it reached the delta of infinite open air, cooled to ash, and disappeared. Then I looked again at the firelit faces of my comrades, a good group of others on the outskirts of certainty. Most of them understood, as I did, that even though comfort and stability were not often among the gifts of wilderness living, there was undeniable satisfaction in the adventure in our work and a kind of security in the company of familiar strangers. And the things we sacrificed—job security, family ties, and permanent homes—were all but forgotten on days like that day when we were paid to swim the cold, clear waters of a mountain stream in search of silver fish.

9

Swimmin' Hole

WENDELL R. HAAG

I am a freshwater mussel ecologist. In streams, freshwater mussels are often found partially buried in the sand and gravel of shallow riffles and runs. Sampling for mussels in these habitats is most effectively conducted by lying in the riffle on your belly and using a mask and snorkel to carefully inspect the stream bottom.

One day I was using this method to sample for mussels in a swift riffle about twelve inches deep in Kentucky's Licking River. While crawling slowly upstream in the riffle with my head in the water, I got that mysterious feeling that someone was watching me. When I raised my head, sure enough, there was a young boy standing on a gravel bar about ten feet away. Not wishing to take off my mask or get up off my belly, I simply waved to him and continued with my sampling.

After a few minutes I glanced up again and noticed that he was still there. "Well," I thought, "this fellow wants to talk." So I sat up in the riffle, took off my mask, and said to the boy, "Hi there."

He looked at me with a puzzled expression and slowly nodded hello without saying anything.

I tried again. "You doin' all right today?"

Again, he returned only the puzzled look and a single, slow nod.

I waited a few more seconds and, not receiving a response, said, "Take care."

Then I put my mask on and started to get down onto my belly to continue sampling. Just as I was about to put my head underwater, the boy said, "Mister?" As I started to rise up out of the water again, he continued, "I know a better place to swim!"

10

The Goat in the Galley

CHRIS SMITH

In 1984 the Alaska legislature passed a bill mandating that the Department of Fish and Game introduce "no less than 50 Roosevelt elk to Southeast Alaska." There was no funding provided in the bill, or any direction on where to get the elk. That was up to us in the department to figure out.

Through calls to states within the range of the Roosevelt subspecies of elk, we determined that Oregon had a surplus on one of its winter ranges. Oregon game officials assured us they could easily round up fifty to seventy-five animals the following winter and would be glad to ship them to Alaska. And the price seemed right; all they wanted in return was ten or fifteen Mountain Goats to release in an Oregon mountain range.

I had been doing research on Mountain Goats for several years out of the Ketchikan office near the southern tip of the Alaska panhandle and had perfected helicopter darting of these rugged animals. Between 1981 and 1984 I had successfully caught more than one hundred goats, losing only three to falls from cliffs after the drugs took effect, so we figured it would be no problem at all to capture enough goats to swap Oregon for the elk decreed for Alaska. I was forgetting, once again, that Murphy's Law exists for the sole purpose of bringing overconfident biologists to their senses.

We planned the goat capture expedition for late July or early August, when the weather and goats were most likely to be cooperative. In Southeast Alaska, where it rains thirteen feet—yes, that's feet—a year, you don't have a lot of opportunities to fly around the top of the mountains in search of goats. As for the goats themselves, they spent only the months of June through September above the forested cliffs where darting from a helicopter was feasible.

You also need to understand that Alaska, and especially Southeast Alaska, is a bit different from most other places. There are no roads. The place is a maze of islands and inlets, its mountains are shrouded in fog and rain most of the time, and just a handful of towns and villages are scattered around at strategic locations for catching and processing the abundant salmon that spawn in the clear waters of Alaska's and British Columbia's rivers. So if you can't fly to where you're going, you have to go by boat.

Given the importance of commercial fishing to the economy, and the need for biologists to spend time out with the fleet, the Alaska Department of Fish and Game has several excellent vessels home-ported throughout Southeast. The main ship in Ketchikan was the *Sundance*, a sixty-eight-footer with the wheelhouse, galley, berths, and a small lab in the aft; a large open deck for storing gear amidships; and a fo'c'sle with additional bunks and storage space in the bow. With more than enough room for the crew of four we'd need to capture and handle the goats, and deck space for over a dozen crated goats, she was the perfect vessel for the job.

We planned to sail the *Sundance* into the head of Boca de Quadra Inlet in the Misty Fjords National Monument and anchor her near an old logging landing we could use as the base of operations. A helicopter from Ketchikan would rendezvous with us there to carry the darting crew and sling the tranquilized goats down to the landing.

Once at the landing, each goat was to be loaded into a crate built from plywood and two-by-fours. The crates were four feet square and about two feet wide with sliding doors on each end. They were a bit big for nannies, which weighed only about one hundred fifty pounds, but had to be large enough to accommodate the biggest billy we might catch; several I'd handled the summer before had tipped the scales at just over three hundred pounds!

On July 29, amid a lull in the salmon fishery and with a forecast of a week of decent weather, the capture crew clambered aboard the *Sundance* and we sailed to Boca de Quadra. The nine-hour run was uneventful. We got our gear ready, inventoried drugs and darts, played a few games of pinochle, and—as always on the *Sundance*—enjoyed wonderful meals prepared by the cook/first mate. We dropped the anchor around ten in the evening on a night as beautiful as anyone could imagine and dreamed of a quick and efficient operation in the days ahead.

As we finished our breakfast the next morning, we heard the high-pitched buzz of the Hughes 500D-model helicopter coming up the inlet. By the time the pilot had refueled, the capture crew was ready to head out for the initial darting run.

As the gunner, I sat in the back door directly behind the pilot. My technician, Vern Beier, sat beside the pilot up front, looking for goats and keeping an eye on any we darted until they went down. When the drugs took effect, the helicopter would drop Vern and me off as close to the goat as possible, and we would climb to wherever the goat was, bundle it in a cargo net, and hook it to a line slung under the helicopter.

The local area management biologist, Bob Wood, and another technician, Charlie Land, stayed on the landing near the *Sundance* to load the goats into the crates and administer the antidote to the capture drug after taking measurements and blood samples.

The first morning's captures went without a hitch. By ten o'clock, when the heat drove the goats into the shade of rocks or timber, we had four goats crated up and onboard the *Sundance*. That evening we caught three more. The following day started slower, with only two goats in the morning, but with the help of the long Alaska summer daylight, by the time it got too dark to fly around eleven at night we had thirteen goats on the deck: six adult nannies, two yearling nannies, one yearling billy, and four billies from two to eight years old. The eight-year-old was a monster!

The weather was holding, so we decided to stay one more night and catch two more goats in the morning. Unfortunately, day broke with heavy cloud cover and a forecast for heavy wind and seas. We sent the helicopter back to town, battened down the crates, weighed anchor, and headed back to Ketchikan with the goats we had. I felt good about getting all the goats we needed for Oregon in two days with no injuries to the crew or goats. I didn't think much about how the number of goats we had on board might affect our luck for the rest of the operation.

The run back to Ketchikan took a bit longer, as we were bucking thirty-five-knot winds and seas to six feet. I'd pull on my rain gear and go out on deck every couple hours to check on our passengers and was pleased to see that animals used to spending their lives on solid, if near vertical, rock handled the swaying motion of the ship better than many of the rookie biologists whose breakfasts had gone over the rail of the *Sundance*. The only ill effect of their confinement seemed to be the accumulating masses of goat turds at the aft end of each crate.

The next phase of the project called for the goats to be transferred to cargo containers and loaded on an Alaska Airlines 737 for a flight to Seattle. Oregon Fish and Wildlife biologists were to meet the plane and haul the goats by truck from SeaTac Airport to the hills of Oregon. With any luck, the first goats caught on the 30th would be running free in their new home after less than seventy-two hours in a crate. With any luck.

Bucking wind and waves, by the time we made it back to Ketchikan, the last Alaska Airlines flight for the day had departed for Seattle. That meant another night in the crates on board the *Sundance* for our increasingly restless captives. We moored the boat in her usual slip in the harbor and draped tarps over the crates, as much to keep the lights and noise of the harbor from spooking the goats as to keep some of the now-driving rain off the crates. It also helped cut down on the number of people who climbed aboard to peek at the odd freight we had on deck.

At dawn the next morning we slipped the lines from the dock and the skipper backed the *Sundance* slowly from her berth for the short run across the narrows to Gravina Island and the Ketchikan airport. The early jet was due in at nine, and we wanted to get the goats on that plane so they'd be released by that evening at the latest.

In an effort to minimize the mess the airline would have to deal with—since it had agreed to donate the cost of the flight—Vern and I decided to wash out the goat crates with a deck hose as we made the short run to Gravina. This chore was expedited by the sliding doors on both ends of the crates. Vern would raise the forward door a couple inches and insert the nozzle of the hose while I raised the back door to let the water—and three days of accumulated goat droppings and urine—flush out the aft.

The plan worked fine for the first six or eight crates. When we got to the one with the eight-year-old billy, our luck changed dramatically.

Although Bob and Charlie swore they had put the billy in head first, and the pile of goat turds at the aft door supported their claim, somehow that huge goat had managed to turn around inside the crate. As Vern put the hose to the forward end and I raised the aft a few inches, suddenly the crack at the bottom of the door was filled with sharp, black horns!

Before I could react, the billy jerked his head upward, thrusting the door open about two and a half feet. I found myself in a squatting position, face to face with the business end of a very angry Mountain Goat that outweighed me by a considerable amount. He'd spent three days in that crate and made it clear in an instant he wasn't happy about it.

I yelled at him—in fear or in a vain effort to frighten him back into the crate, I'm not sure which—and lunged for the top of the door, hoping I could slam it shut before he bolted. However, when he had thrust the door open with his horns, he had jammed it in the frame, and that which slid up so easily would not go back down in spite of my one hundred ninety pounds pressing on it.

If you picture the head and horns of the goat on the outside of the sliding door, and me trying to push the door back down, you get some idea of the proximity of his ten-inch horns and some fairly delicate—and important—parts of my anatomy. Needless to say, when the door didn't budge after the first couple attempts, I let discretion—and the hope for future children—prevail and bailed off the crate. The billy made his move and in a flash was free on board the deck of the *Sundance*.

I don't recall how long he took to get his bearings, but it wasn't long. After looking left and right and seeing his way blocked by crates or other gear on the deck, he trotted down the gangway along the starboard rail, as if he'd been sailing all his life.

I've been told Mountain Goats can swim, though after watching hundreds of goats for years I have never seen one in the water. He must not have believed goats could swim either, because in spite of having ample opportunity to escape over the rail and into the harbor, he apparently never even gave going overboard a second thought. True to his breeding, he was looking for a cliff!

At the aft end of the gangway, he had only three choices: turn around and come back at me and the rest of the crew that had gathered in response to my hollering; go overboard, which as I said, he'd already considered and abandoned; or turn right and go through the doorway into the galley. Don't ask me why the door was open; ask Murphy—he's the one who left it that way.

The goat chose the door.

Anyone who's been on a boat knows the only thing more cramped is the inside of the space shuttle. Even on a sixty-eight-foot boat capable of carrying a crew of twelve, the galley is a tight space. The stove is on one side and the sink is on the other with about twenty inches of counter space between. It's a space meant for the cook, and only the cook, to work in. In addition, there's a table that will seat four at a time and the steep steps to the wheelhouse.

By the time the crew and I had followed the goat along the gangway to the galley door, he had reconnoitered the place and decided the counter between the stove and sink would be his Alamo. It probably looked like a small ledge on a cliff to him, gave him a place somewhat higher to work from, and put his back to a wall so that whoever came after him would have to deal with the two ten-inch daggers sprouting from the fore end.

We were relieved, actually, that he had chosen the counter. Had he taken the steep steps to the wheelhouse, he'd have had command of the ship. There being no door at the top of the steps, when the skipper saw the goat come into the galley, he threw the engines into neutral, abandoned the bridge through the back door onto the poop deck and left us all adrift in the midst of Tongass Narrows. I don't know if the goat could really have piloted the *Sundance*, but I'm sure his horns would have taken a toll on the ship's expensive electronics before it was all over.

As the four of us on the crew peered through the door and galley window, the billy kicked four days' worth of dirty dishes from the counter and sink, backed his hindquarters into the corner, lowered his head, and seemed to dare us to enter the confined space. Three of us were smart enough not to take the challenge.

Charlie Land was a strapping former Texan who'd worked on a private game ranch there. He was used to wrestling with all sorts of exotic animals and was not about to be intimidated by a mere Mountain

Goat. Even one that was clearly a prime specimen, extremely pissed off, and in a good defensive position. Before we could talk him out of it, Charlie waded into the galley with intent to subdue the goat before he did any more damage or caused us to miss connections with the morning jet.

When Charlie was still a couple feet away from the goat, the billy dropped his horns and launched off the counter. In a move driven purely by survival instinct, Charlie caught the goat's twin rapiers just before they entered his gut and fell back on the floor between the stove and the sink. From the safety of the door, all we could see was a couple hundred pounds of thrashing Mountain Goat with two human legs sticking straight up in the air alongside it. The air was filled with a cacophony of shattering dishes, clattering pots and pans, and profanity uttered in a boisterous Texas drawl, several octaves above Charlie's normal tone.

Given that Charlie had the goat's horns in either his hands or his abdomen, the rest of us figured it was safe to join the fray. Although the goat was strong, with three grown men on top of him and one underneath with a death grip on his head, we managed to subdue him fairly quickly. That is to say, once we got enough biomass jammed between the stove and the sink, not even an angry goat had any wiggle room left.

After taking a few seconds to make sure Charlie hadn't been impaled, we began trying to figure out what our next move was. Amid the heavy breathing and ongoing occasional curse, Charlie told us that when they'd handled recalcitrant critters like this on the ranch in Texas, two guys would take one hind leg apiece, one guy would take one front leg, and the fourth guy would handle the head. That way, the animal would have only one leg to hop on and could be manhandled however the crew wanted. The Texans must have been stronger than we were.

When we tried Charlie's plan, all that happened was that Vern and I, who drew the back legs, got pummeled by hind hooves the size of a linebacker's fist, driven by thighs the size of...well, a linebacker's. With no room to stretch the goat out in the galley, we couldn't gain any physical advantage over him. The only benefit of our second round with the goat was that we managed to get Charlie out from under him.

After a few more minutes of brief conversation, and realizing we needed to do something before the *Sundance* drifted onto the rocks, I opted for drugging the goat. While the others held him down, I ran to the fo'c'sle for some M-99 and a syringe. Five minutes later the goat was back asleep. In five more he was in a new crate—with both doors firmly latched. The skipper regained command of the bridge, and we docked in time to get all the goats onto the jet. When the cargo crew weighed the crate with the billy, it tipped the scale at 372 pounds. Subtracting the weight of the crate meant the billy came in at 298 pounds.

As the plane roared off to the south, the *Sundance* made her way back to the harbor and Vern and I helped the cook clean up the galley. Aside from the broken dishes, a saucepan that looked more like a crushed pop can, and a few bruises, there wasn't much evidence of the morning's adventure. But those of us on board the *Sundance* that day won't soon forget the goat in the galley.

II

A Fox's Love Story

RAN LEVY-YAMAMORI

It happened one summer on the island of Hokkaido, the most northern island of Japan. Winters in Hokkaido are very long and cold—
everything freezes. But during the brief, warm summers everything
turns green and blooms in a blaze of color. That particular summer I
was collecting information on plants for a book I was writing. Since
the trip was long, my friend Itai joined me to help with the driving. In
an old decrepit van, which also served as our crude sleeping quarters,
we set out in search of wildflowers.

One evening, after a long and exhausting hike, we drove toward
the seashore along a route that appeared, on the map, to be a shortcut.
It turned out to be a narrow and bumpy road that meandered through
cultivated fields and conifer forests. I was driving while Itai napped in
the seat next to me. We still had a long way yet to drive when darkness
settled on the hills and the curving road. The place was so remote that
I saw neither lights nor any other sign of human settlement. I imagined
that there was nobody else in the world except for us and the insects
that were attracted to the headlights.

Humming a song to stay awake, I crossed a small bridge, and then
suddenly the moon appeared between the clouds, turning everything
pale blue. In the ethereal light I noticed something lying on the road
ahead. I slowed down, thinking it must be a dog, but as I approached I
realized it was too small. Perhaps a cat? No, the ears seemed too large.
"Ahhh...a fox!" I concluded and stopped the van.

I really do not know why I stopped. After all, it was only the carcass of a fox, similar to many I had seen before. But I pulled the van to the side of the road and turned off the engine, leaving the headlights on.

I opened the window to let in the chilly air and welcomed a concert of croaking frogs, chirping crickets, and the occasional calling owl. Itai woke up, wondering why I had stopped. Before I could explain, we both noticed a pair of eyes sparkling at the side of the road. We froze in our seats so as not to scare the hesitant animal.

The pair of eyes came a little closer and then halted. The animal took one more tentative step, and we recognized a silhouette lit by the margins of the headlights.

"A fox," I whispered, observing its narrow tail, which lacked the male's typical flag-like white tip. "It's a female."

She looked at us and then turned to the dead fox in the road. Cautiously, she looked again at us, trying to overcome her fear. Another small step, and she stopped to check our response. Throughout those seemingly endless moments Itai and I sat still, afraid to move or even to breathe. The fox kept her eyes on us while slowly, step by step, she approached the dead fox. With her eyes still fixed on us, she sniffed it, then uttered a soft sobbing sound. She grasped the neck of the dead fox between her jaws. Stooping from the effort, she carried it to the side of the road and slowly disappeared into the darkness.

Since the moment I realized I was to become a naturalist, it was obvious to me that when observing wildlife, one must separate sentiment from scientific observations. In my scientific work, therefore, I have always tried not to interpret animal behavior through any human way of thinking. I have never been one to say "this animal loves" or "that animal hates." But watching the fox drag her mate from the road, I could not control my emotions. My eyes welled with tears that rolled down my cheeks. I peeked at Itai to see if he noticed my reaction, and for a while neither of us spoke. Then Itai said something that I, the naturalist, did not dare say: "What a brave fox!" he cried. "What love!"

12

Fire: A Prairie's Own Companion

ALICE CASCORBI

October 1991. I was chasing a fire across the Bluestem Prairie, a two-thousand-acre tract of northern Minnesota's Red River Valley, sixty pounds of water sloshing in a brass can on my back. The fire was ahead of me, moving through the grass like a leisurely rhino—slow right now, but nothing to turn my back on. Warily, I followed with my water. The wind gusted in my face. Burning embers rained down all over, finding fertile footing in the dry autumn grass. As afternoon temperatures climbed into the eighties, I charged from one flare-up to another, dousing them with a hand pump connected to the water can. The load on my back grew lighter as the hours went by but felt heavier as I grew more exhausted. I lost sight of the flames in the billowing smoke that hung a pall over this near-flat Minnesota landscape. When word came through toward sunset that the fire was finally out, I was elated. I met up with my colleagues, seven smoke-stained people in yellow fireproof coveralls, and we shared weary high-fives. This time we would get home before midnight.

We walked over a landscape that an hour ago had been thick with life, lush with grasses and tall goldenrods. Now, around us and underfoot, the ground was bare of all but stubs of vegetation. Black ashes of the grasses stood in place where the fire had overtaken them—fragile, carbonized replicas of their living forms, like the powdery fur of some mummified animal. Burrows of mice and ground squirrels lay revealed as smoking holes, now that fire had stripped away their cover. It seemed a scene of desolation, but we were pleased—and not just because our hard day's work was done. We had not come to fight the fire; we were the team that started it. This land belonged to the Nature Conservancy. We were burning the grassland to restore its health, and today had been an easy, thorough burn.

The first time I saw prairie, I would never have believed I would come to love it enough to take that dirty, sweaty, low-paying job as a burn crew intern. I also might not have believed I'd feel rewarded by seeing a lot of it go up in flames. I grew up in the loamy green of Ohio's woodlands. All I knew of prairie, before I went to college in Minnesota, was gleaned from casually thumbing through Sierra Club calendars. I didn't know where they'd taken the August photos of prairie, blazing with wildflowers. And I had even less idea that this celebrated ecosystem had an intimate relationship with fire.

As it turned out, my college tended a prairie remnant. And when I started college, I couldn't wait to see it, couldn't wait to stand waist-deep in nodding flowers. But when a senior biology major proudly showed it to me, I saw ten acres of browned grass under the September sun, and I was sorely disappointed. Where were the Technicolor flowers—the colors that put prairie on the calendar? I couldn't tell this field of withered grasses from any common hayfield. The enthusiasm of the upperclassman talking about "native forbs" and "bison-adapted ecosystem" didn't stir me. Privately, I thought that prairie, like many other things you see on calendars, must just not be as beautiful in real life.

I began to change my mind as I began to bend down more. A prairie is not immediately spectacular, especially if your eyes are attuned to forests. In a prairie most of the action happens at knee level. If you can bend down, spend time eye-to-eye with the vegetation, you will begin to see its beauty. If you can't find the bison, watch the insects, birds, and rodents that make the grass their home—you will see the richness of a rain forest, on a miniature scale. Wildflowers do come, in their seasons; the prairie is always changing as sunflowers, clovers, asters, and goldenrods come into bloom and fade in their time. But the textures of the grasses themselves, the complex interplay of more than forty species, give the land a quality beyond the bright hues of the flowers.

Three types of prairie historically graded into one another across central North America: shortgrass, midgrass, and tallgrass or "true" prairie. Shortgrass prairie, a steppe landscape with grasses about eight inches tall, defined the Great Plains from the southern Rockies into western Nebraska, Kansas, Oklahoma, and Texas. Eastward, this steppe graded into midgrass prairie, twelve to eighteen inches tall, which was found throughout most of Montana, the Dakotas, Saskatchewan, and Manitoba and ran in a narrow band southward into Texas. And finally, from the southern part of Manitoba, tallgrass prairie cascaded down to the southern tip of Texas, reaching as far east as Ohio in the widest part of its range. This was the region where grass could tower above a horse's ears, where wagons moving westward truly were prairie schooners, afloat in a sea of grasses.

Fire is especially vital to the tallgrass prairie. On the Great Plains and in the midgrass region, too little rain falls to let trees grow. But in the eastern portion of the native grassland ecosystem, there is enough rain to support either woodland or prairie. It is fire that keeps the forests back. Working on the burn crew for the Nature Conservancy, I saw first-hand what this had meant for the land. The ashes of the grasses, crushed underfoot, turned the soil the exact black color of the new-plowed earth in spring. The topsoil here counts among the most fertile in the world, and it got that way because of sod and fire.

Sod is the prairie's thick mat of interlocking roots. It repels not only most invading weeds but even farm implements made of iron. Before the invention of the steel plow, settlers could not break the sod, and there were cities established on both coasts before prairie exploitation began in earnest. When John Deere invented the steel moldboard plow in 1837, the sod could be peeled back to reveal black topsoil more than twenty feet deep—built up by grass and fire in the ten thousand years since glaciers last swept the land clean. The settlers plowed the treeless land and set about fulfilling a government mandate to "refor-est" it, planting spindly sticks of willow and box elder to mark fencerows. In thirty years the prairie was reduced to remnants, pushed onto tiny reservations where the plow is banned by geology or custom: steep hillsides, railroad edges, graveyards. Today less than 1 percent of America's tallgrass prairie is still intact. Almost all of it is now our grain belt, growing crops such as corn and soybeans.

To the settlers with their dreams of prosperous farms, prairie fire became a nightmare. Working on the burn crew, I learned why. Sometimes, when the headfire is set and it's traveling with a good, stiff wind behind it, the flames really do leap up twenty feet, and the wall of smoke and flames tearing down the prairie is terrible and awesome. I can understand the fear of the farmers, rooted in one spot to their barns and buildings. Fire control became a rallying point of pioneer settlement. Fire lines were plowed, and windbreaks were planted. The land became a patchwork quilt of disconnected parcels, sown with European grasses and edged with non-native trees.

But fire had always been part of the prairie system. Thunderstorms spawned lighting fires, and native peoples, taking a cue from nature, would deliberately fire the grass to clear the way for travel and rejuvenate range for their game and horses. Their mobile lifestyle allowed them to work with prairie fire instead of against it.

On the Nature Conservancy's burn crew we applied fire's power with due caution. Our fires were set in a ring; that way, the sides burn toward each other, meet in the middle, and go out. But a circle fire is still an awesome sight. Smoke from two sides begins to rotate around in a fire tornado, a gyre that pumps smoke hundreds of feet into the sky and spins off twenty-foot flaming dust devils around its base. Wind hits your face, loaded with the intoxicating smell of burning sage, with the bitter, satisfying fragrance of woody brush giving up its life into the atmosphere.

Our burns were choreographed with minute attention to wind, humidity, and the day's expected temperature so that wind would carry the flames from all sides almost to the middle of the burn site before the fires met and got this violent. From a safe distance one can watch it as a spectacle, like wild bulls in a rodeo. But fire was something we wrangled and corralled, not a force that we domesticated.

To live with fire, native plants dig in for the long haul. The roots of prairie plants can run nine feet deep; both grasses and wildflowers keep the bulk of their tissue underground. Most prairie plants are perennials. They spend their first growing season as tiny shoots, investing their growth energy in their root systems. Only when this underground safety net is complete do they soar up and work toward flowering. Like a cellar full of preserves, this investment in roots lets native plants survive a fire and come back strong. When a blaze rips through, it kills only the flimsy top of the plant. The underground portions are safe, ready to spring up again.

This is the crucial difference between native prairie plants and the European plants now competing with them. European grasses, planted by settlers, evolved in cooler meadows without fire. They have shallow root systems, and many are annuals. European weeds, such as wild radish, spring up tall their first year and spread thousands of seeds. Disturbing the soil, by plowing or intensive grazing, gives these plants the upper hand. Their tall stems shade out the slow-growing seedlings of prairie plants, and they set massive amounts of seed to do the same thing next year. But the European species cannot withstand a thorough cooking. Regular fires destroy them, leaving the field clear for fire-adapted natives. Within small remnant prairies many native plant species dwindle to extinction, but burned remnants lose fewer species than unburned remnants. Fire helps preserve native plant diversity.

Strangely enough, fire does most of its beneficial work by letting in the sun. A burn is a kind of tonic for the soil. The upper few inches of prairie earth are alive with microorganisms—bacteria and fungi—which love the warmth as the sun shines down on soil cleansed of clogging debris. The microorganisms grow explosively, pumping nutrients into the soil that feed the roots of grasses and wildflowers. The warmed soil also benefits the plants directly: most prairie grasses grow best when temperatures are high. By removing the insulating thatch layer, fire extends their growing season, letting them gain more than they lose to the flames.

Prairie animals, too, have adapted to fire. Bigger animals, such as elk and bison, can run clear of most fires, and smaller animals, such as reptiles and rodents, dive into burrows to let the flames pass by. Of course, some casualties do occur—baby mice and birds can die if fire finds them in aboveground nests, old or weak animals may not be able to escape a fast-moving wall of flames, and as pioneers recorded, buffalo and horses sometimes perished when surrounded by fire. But by and large, prairie animals have learned to live with fire. My favorite example of this is the prairie grasshopper. These grasshoppers use every natural color of the prairie for camouflage. One species is the yellow-green of new grass growth, one the olive of old goldenrod stalks, another the soft light brown of last year's hay. One grasshopper is the exact gray-green of prairie sage; another is—surprise!—the shocking pink of coneflower petals. And in this insect rainbow there is even a grasshopper the exact charcoal brown of scorched grass stubs. When "char" is a color that a species can adopt as camouflage, it is evident that fire is an integral part of its ecosystem.

I went back to Bluestem Prairie the year after I'd helped burn it. The thick, healthy stands of green bunchgrass showed that we had done our job well. The charred remnants of non-native box elders were still in place, surrounded by the purple native blazing-star just coming into bloom. Looking out over waving fields of flowers that have sprung up after a fire, you can come to understand why the Hindus consider one god, Shiva, the divine force behind both destruction and creation. Fire re-creates the prairie. In returning it as a management tool to the small islands of native grassland we have left, we give this beautiful ecosystem a second chance to live.

13

Road Hunter

DAVID M. LIBERTY

Hunting season brought all the crazies into the field. As Forest Service employees on the Malheur National Forest, we encountered the worst road hogs and the laziest road hunters and were asked a thousand times where all the deer/elk/fish/birds/women were, as if we were actually going to tell anybody. Nothing really shocked us. We thought we had seen just about everything until the year of the coat in the road…

That particular deer season my co-worker, Greg, and I were driving up the Middle Fork of the John Day River in beautiful eastern Oregon when we saw a blue coat lying in the middle of the road. We suspected someone had left it on his roof and had driven out from under it. Greg got out to see if it was worth picking up when I saw his eyes get really big. I looked in the direction of his concern and saw a leg sticking straight up, hooked to a barbed-wire fence down off the toe of the road fill. We both thought we had found a dead body! Every year it seems like at least one hunter ends up dead from a stray, or well-intended, bullet.

I got out, and as we crept closer, the body began moving. The fence-tied figure jumped up as if to pretend there was some legitimate reason for him to be hung up on a fence unmoving until we arrived on the scene.

"I'm huntin'," he said. There was no sign of a gun. "My buddies dropped me off here 'n' said they'd come get me, but that was 'bout four hours ago."

By this time it was apparent the guy had over-imbibed at the bar back in camp and had probably forgotten how to negotiate barbed wire when the fence reached out, wrestled him to the ground, and knocked him out.

"Where's your camp? We'll give you a lift," we offered, being the good Forest Service hosts we were. We didn't want to read in next week's *Blue Mountain Eagle* about a blue-coated drunk found drowned in the Middle Fork.

"Sure," he slurred, picking up his coat. "I think my camp is jussup river."

We figured he must have been a west-side hunter. They are the bane of our existence as they migrate full force to our side of the state every fall for mass murder, unsupervised drunkenness, reckless littering, and general onslaught.

"Hop up on the hood. You can hang on to the hood latch," we offered, knowing a little wind in the face can help clear the head in cases such as this and surmising that his breath would probably reflect the effects of a four-day drunk with no bath or even a toothbrush. Besides all that, it is illegal for civilians to ride in a Forest Service vehicle without special permission from the president himself. We proceeded upriver until he waved at us to stop near a road junction where a big blue Ford pickup was parked. He waved to his buddies, who looked at us as if we had brought the ex-wife's mother-in-law back from the dead. Not wanting to encourage any friendly (or unfriendly) gestures, we left immediately, patting ourselves on the back for saving another life.

We headed on into town, anxious to share our "you'll never believe this" story with our friends, family, and co-workers, making a mental note to ignore all coats in the road until hunting season was over.

14

Snares

TROY DAVIS

I glanced back over my shoulder. The ground fell away, sloping steeply down to the minor drainage where I had stood what seemed like hours ago. I looked up. From my feet, the ground surged upward toward the sky, where a lone, gnarled pine tree stood on the horizon. *That damn tree is laughing at me*, I thought. And although I had no plausible explanation as to how, it was also steadily moving farther away.

As if the realization that I was angry at a conifer wasn't enough, a small figure just below the tree caused my spirits to sink even more. It was Mike, my field partner. His tiny silhouette appeared relaxed and patient as he periodically peered back. Far below him, I was carrying a load of psychological distress in addition to my field pack, both of which weighed heavily on me as I followed his tracks in the snow.

I'm not going to make it, I thought with every third step. All things considered, it could have been worse. Death on a beautiful, snow-dusted Montana ridgeline, bordered by the Absaroka Range and Gallatin Mountains, was at the very least an epitaph I could be proud of.

As I pondered how I came to this premature, high-altitude demise, my mind drifted back to the events of the preceding months.

That winter, I had accepted a position as the lead technician on a winter wildlife study in Yellowstone National Park. My job was to manage incoming data and generally support the field crews, two things that placed me in front of a computer screen for more time than I had ever imagined.

In the ten years of work preceding this assignment, I had spent the majority of my time in the field. I drove old four-by-four pickups with holes in the floor. I carried ridiculously heavy backpacks and awkward GPS units into places that no person in their right mind would go. I slept under trees, in trees, and perched on an ATV. I had lived in seven states in ten years, driving back and forth across the country for jobs that sometimes lasted only a few months. I did these things not because I was enamored with vagrancy but because I cared about the animals I worked with—species ranging from moths and sea turtles to spotted owls and bison. In fact, I had become a biologist because I loved animals. It was, naively, that simple.

Suddenly, with precious little warning, I had become an office biologist. I sent out and replied to obnoxious volumes of e-mail. I managed databases. I set up meetings and conferences (although I was never sure how a *meeting* differed from a *conference*, or a get-together, for that matter).

The realization itself wouldn't have been so bad, were I an effective and efficient administrator, but I was not. I have bad handwriting, I couldn't concentrate during planning sessions, I couldn't read legalese, and it was by the slimmest of margins that I even managed to get time sheets turned in. I lacked the social skills to write evenhanded, polite letters thanking people for their interest in complaining about too many wolves eating all the elk, or too many elk eating all the grass, or too many snowmobiles going too fast, or that the speed limit was too low.

Granted, I was involved with an important project with potentially significant consequences for the most famous national park in the world, and I worked for intelligent and influential scientists. But that legitimacy, somewhat unexpectedly, weighed heavily on me. In my seasonal position I orbited far away from the centers of power in the Park Service hierarchy: if this situation could cause me such distress, what would higher career paths hold? Many of the people I worked for had been promoted out of the field only to find themselves sitting in work-group meetings, drafting proposals, and balancing budgets. Was it foolish to believe that I could spend my life with snakes and bears, or was budget training and an intimate knowledge of the National Environmental Policy Act the only future I could look forward to?

The perennial workaholic in me reasoned that the only way to deal with this confusion was to spend so much time working that exhaustion would preclude my worrying about it. I went out several times to check on the field crews to distract myself from growing doubts. This consisted of riding a snowmobile for two hours to track down all the competent and experienced field people on the study crew and inevitably recognizing that they all knew exactly what they were doing and consequently had very little use for me.

As part of my clever plan to exhaust my anxiety, I had contacted Mike's boss and indicated that I wanted to get in on some of their Mountain Lion work. Of all the species I had seen, worked with, and been trampled by, I had never worked with cougars. The project was in the early stages of laying out an intensive hair-snare grid, and Mike was happy to have some volunteer help.

One February morning we towed an aging snowmobile to the end of the road above a tiny town in Montana. We haphazardly strapped our gear to the already awkward contraption and sped off, continuing our journey up the mountain. Reaching our destination, we donned our snowshoes and headed into rugged terrain carved out by the Yellowstone River below.

At first we waded through deep snow, postholing across wind-crusted flats and sparsely treed ridges. As we entered the more forested environs of the river canyon, the going became easier. Protected from harvest for over one hundred years by the park and by the steepness of the landscape, the ancient conifers grew thick boughs that prevented deep accumulations of snow. Eventually, the lack of snow and the preponderance of rocks and downfallen logs persuaded us to remove our ungainly snowshoes and trek on in our Pac boots.

Small, winter-weary groups of cow elk flushed from the dense pine stands as we hiked. Bulls, roused from their daybeds, snorted in irritation and then vanished, silent and ghostlike, into the trees. Amid the boulders we found the scattered, weathered bones of long-forgotten kills. It was perfect cat country: rocky, beautiful, and harsh. The ground itself seemed to invoke the American Lion.

Mike was setting up three separate hair-snare grids that day. A concerned young girl once asked a lynx researcher what he did with the rabbits (hares) after he snared them. He explained that hair snares catch exactly that—hair. The snares are simple arrangements consisting of a visual lure (a shiny pie plate hung to spin in the wind), a scent lure (an odiferous mixture of catnip and beaver castoreum), and a number of small, scent-impregnated fabric squares supporting a phalanx of studded spikes that are attached to a tree base. When an animal is attracted by one of the lures, it approaches, localizes the scent in the fabric squares, and then responds to the perceived territorial incursion by rubbing its own scent over the beaver's. As it rubs against the fabric squares, a few of its hairs are snagged on the spikes. Some of those hairs have intact roots that can yield DNA, which laboratories can then use to identify the species and, in some cases, individuals.

This is, of course, a theoretically perfect and antiseptic scenario. In reality, animals can refuse to respond to the lures. Elk can respond to the visual lure and treat the pie plate as bull elk treat most everything—by fighting with it until it eventually resembles shredded tin foil. In the spring and summer, bears can be attracted to the scent lure and do what bears do best—demolish the entire site. Animals can visit the lures and decide against donating any hair samples. Even if the stations survive, the animals come and rub, and the samples are gathered with the utmost care, mishaps can occur in the lab. Such samples can be extremely important and even controversial when involving lynx or wolverine, species whose presence can impact land management decisions. Pressure and rhetoric can run high.

Mike's project concerned the slightly less contentious cougar. Although lions are not beloved by any means in the western United States, they are grudgingly tolerated in many states as a trophy game animal. His grid was intended to detect and approximate the number of individual Mountain Lions in the study area.

In the process of laying out the snare grids, we gained and lost elevation, stumbled through talus fields, and gripped withered plants to drag ourselves up slopes. We clawed our way along exfoliating rock ledges. In retrospect, it would be more honest to say that *I* did most of these things. *I* knocked my knees against rocks. *I* stumbled and fell on every available sharp branch. *I* wheezed like a leaking balloon. For his part, Mike glided smoothly across the rough terrain and pretended not to notice my flailing.

I had feared that the months of office work might have reduced my fitness, so I had begun running nine miles a week and lifting weights. All my effort, I understood that afternoon with a painful immediacy, had done little to help me prepare for chasing cougar. Still, I kept up with Mike at a respectable distance for most of the day, despite my complaining body.

Ironically, it was the final leg, the home stretch, where I finally faltered. We had placed all the snare grids successfully, and only one long ascent remained between us and the snowmobile. The rest of my body was grudgingly cooperating, but the long uphill climb caused my quads to mutiny. I grew numb from my hips to my knees. Mike was slowly disappearing as I fell farther behind.

I wouldn't give in; I wouldn't stop—I *couldn't*. Even if that were the reasonable and prudent thing to do, I couldn't. I had something to prove; I could not afford to be reasonable. I had shown little aptitude behind a desk, so I had to prove to myself that I still belonged in the field. To make matters worse, I had harbored a schoolboy crush on Mike's boss, one of the preeminent lion researchers in the country, since I was twenty-two years old. For the better part of a decade, I had persistently pestered her for an opportunity to work lions. She had politely and just as consistently refused until Mike's project opened up space for volunteers.

Now, after the years of petitioning, fate had intervened and granted me one chance to prove myself. If I failed, I would disappoint not just Mike and myself but someone whom I had quietly admired since college. I couldn't afford to fail—emotionally, physically, or professionally. I reflected with somber acceptance that an honorable death on a ridge was quite preferable to that ignominy.

I looked upslope at Mike and that damn evil tree. I made a decision to make some progress before my coronary arteries took me to greet the hereafter, and so I took ten steps. Then I stopped for ten seconds to lean on a broken branch that was I was using more like a crutch the higher I went. Another ten steps, another ten seconds of rest. Neither speed nor grace was my ally on that ridge—pure obstinacy carried me, distilled belligerence borne on a wavering ego. *One foot*, I mumbled to my thighs, *in front of the other*.

That logic—as dull and mechanical as it was—served me surprisingly well, despite my repeated declarations that I was not going to make it. After an embarrassing amount of time I topped the ridge and recognized the spot where we had left the snowmobile. Mike was already there, sitting on the machine. I watched suspiciously for any sign that he was checking his watch to calculate my lag, but he never did.

"Man!" he laughed. "That was the worst transect I've ever done." He displayed a big, infectious smile. "I'm exhausted. How about you?"

I eased myself over to unstrap my snowshoes, testing for any twinges in my back. "I can understand how that might make someone tired," I replied noncommittally.

"The crew is taking tomorrow off," Mike said as he coaxed the snowmobile to life. "Want to come out again on Monday?"

Do it again? I stood dumbly for a moment, listening to the sputtering of the machine, snowshoes dangling in my hand. After the ups and the downs, the bruises and the blisters, the real test had suddenly, unexpectedly presented itself.

Do it again, indeed.

"Sure," I said as I climbed onto the snowmobile. "Count me in."

We raced down the mountain. As I set my pack in my lap and buried my face in the nylon to avoid the cold wind, I reflected on the day and how it boded for my future. I still wasn't sure where my career was headed, but I knew I could contend with any path it presented. I just had to keep putting one foot in front of the other.

15

Pileated Woodpeckers

PATRICK LOAFMAN

There were four of us: my brothers Mark and James, our friend Dave, and me. Night was folding in on us, and we didn't have a tent to sleep in. We had planned to sleep in a shelter—a lean-to that the map said was in this valley—but all we found was a pile of rotten wood where it once stood.

We drew straws. Mark and I drew the shortest ones, so we had to venture back in the oncoming darkness for the tent. We were not far down the trail, so we made it back to the car before it turned pitch-black. We were in the Smokey Mountains. It was summer. A park ranger came by.

"You're not going to hike in the night, are you?" he asked.

We showed him our lantern, and that eased his concern. After he left, Mark tried to light the lantern, but the mantle dissolved to dust. We had no flashlights, only a box of matches. A light misty rain began falling. We couldn't sleep at the car, leaving the others down in the valley without a tent in the rain.

We ventured into the night, into darkness more dense than I had ever been in before. Mark would light a match, giving a small glimpse of the trail for a moment, then the match would burn out, and we would stumble and curse. The small light from the match, that brief glimpse of the world as we knew it—the world of daylight and sight—disappeared all too quickly. How long we stumbled down that path, I cannot remember. I'm sure it seemed longer than it actually was.

We eventually ran out of matches. We reverted to crawling on hands and knees in the mud, feeling for the trail like blind men, inching our way through the starless night. We finally made it back to James and Dave. We made it through the darkness, pushed back its thickness and its fears.

A decade later, after Mark's death, I remembered our success—how we proved to ourselves that it was possible to get through such darkness.

My introduction to radiotelemetry involved Pileated Woodpeckers in the Olympic Mountains of northwestern Washington. The Pileated Woodpecker is a robust, redheaded comedian of old-growth forests. The Olympic woodpeckers were fitted with radio transmitters held in small nylon backpacks that fit over the birds' "shoulders." I held one of these birds, a male, as the backpack was put on. He hammered his bill into my hand, drawing blood as bright red as his crown.

Two weeks later I crawled on my hands and knees with a radio receiver, tracking a beeping signal through a thicket of young Doug firs. I found the woodpecker tapping on a stump. I crawled closer and closer, concealed in the thick branches, and then I turned off the radio receiver and inched closer on my belly like a snake. The bird was a mere foot off the ground. I could see ants crawling from the hole it had dug, and the woodpecker flicked its tongue out, lapping up the insects. I knew it was a male because of his red mustache stripe that shone brightly in the sun.

He stopped eating ants for a second, turned his head as if to look down at me with his yellowish eye, then drummed a few more times on the stump, scattering wood chips. This was the same bird I had held two weeks earlier, the one who had hammered my hand with his beak. I still had a small wound, a star-like scar at the base of my thumb.

I inched closer until I could have reached out and petted his back. The bird guides claim that Pileated Woodpeckers are sixteen inches in length, but this bird seemed much larger, gigantic in stature. I wanted to touch him, to hold him again in my hands.

I had been lying there only a few seconds, holding my breath, before he raised his red crown and turned, and then I heard the wind pushed by his wings as he took off. I turned on the radio receiver and listened as the beep got quieter and more distant.

I knelt in front of the stump and saw a black ant crawl from the woodpecker's hole. I put my face next to the stump, stuck out my tongue, caught the ant, and swallowed it down.

Later I felt foolish.

Pileated Woodpeckers not only nest in cavities; they also roost there, sleeping at night in the centers of trees that are hundreds of years old. Both the male and female spend time on the nest, sharing the chore of incubating the eggs. The male usually takes the night shift while the female spends the night in another cavity. She doesn't stay in the same cavity every night, though; she tends to move to a different cavity every five days. These birds know their wilderness territories as well as we know our neighborhoods and towns. They know exactly where all their old cavities are; some are old nests.

"It's crucial to find where the roost trees are," the project leader maintained. She was a forty-year-old woman with hair as red as the woodpeckers themselves. "This will broaden our understanding of habitat requirements. To find these roost trees, though, requires hiking at night. Are there any volunteers?"

I remembered how Mark and I had stumbled through the Smokey Mountain darkness so many years ago. I raised my hand and volunteered for the night work. I didn't know why at the time, but I needed to go, to push into that darkness once more.

I went out in the evening and drove around the gravel logging roads with the radio receiver on until I found the place where the beeping signal seemed the loudest, and then I waited. As the sun set, the signal moved as the female woodpecker flew to her roost cavity, and it didn't move again. She was settled for the night.

Wrapped in nylon raingear, wielding a flashlight, I stepped off the road into the forest and followed the beep wherever it would lead me.

Many people fear the darkness and all it hides, and indeed I had momentary fits of fear as I splashed through creeks, thinking I heard footsteps somewhere behind me. Sometimes I'd turn my flashlight like a gun on the night and shout, "Who's following me?"

Did I expect to find some demon of mine in the shape of a cougar or bear? Did I expect to see Mark striking matches, illuminating a small circle around him, urging me to continue? I don't know. All I saw in my flashlight, though, were devil's club and salmonberry surrounding me in thorny silence.

Around midnight I realized I was following a bounce, not the real signal. Radio signals bounce around in the mountainous landscape, making this tracking technique an art. It's not always clear which way the beep is louder; it's easy to turn the wrong way, thinking you are following the true signal. So there I was in the middle of the night, nowhere near the bird. I felt, in the purest sense, what it means to be lost. I wondered once again if the cougar was there, just beyond the reach of my light, waiting for its chance to pounce on me.

I shouted at the darkness.

I was too tired, too wet, and too cold to continue searching. I gave up. I retreated back upslope, finding my truck. With a turn of the key the engine came to life, the world lit up, the heater blew warmth, music sang from the radio, and I felt light-years from that cold moment in the dark.

The next night, though, I returned, and this time I found the tree where the female woodpecker was roosting.

I walked around in circles, making sure the beep was coming from the tree. Then I turned off my light, placed my hand on the bark of the tree, and tried to feel her presence somewhere in there, sleeping. I was not afraid. I had company, even if it was merely a sleeping bird two hundred feet above me. Even more comforting was my feeling of success, of venturing through the dark to the source. For one fleeting moment in the dark, touching the bark of the tree as if taking its pulse, I felt a small flash of peace, like a flicker from a match briefly illuminating the trail.

Maybe some of us have to make peace with the darkness, to venture into uncertain landscapes, searching for something familiar. Maybe some of us have to step off into the darkest of nights into the cold, misty rain and wrestle with shadows that follow us. Maybe we must physically take on those shadows, force them to the ground under our own strength, and hold them close enough to smell their breath so we can stare them in the eye and know what's scratching beyond our small circle of light. Others who do not have these needs will mistakenly see what we do as bravery.

16

Museum Sisters of Cheerful Disposition

BARBARA BLANCHARD DEWOLFE

In the early 1930s Mary Erickson and I were the only women teaching assistants in the Museum of Vertebrate Zoology at the University of California, Santa Barbara. During the spring of 1935 Mary and I traveled to northern California to collect pocket gophers for the museum. We were instructed by Joseph Grinnell, the museum's director, to collect five female gophers, plus whatever males were trapped with them, at each point on the map he had prepared. He also gave us a general letter of introduction to farmers who might want to know why we wished to trap gophers on their land. In this letter he stated, "These women are of cheerful disposition."

From then on, Mary and I called ourselves the Museum Sisters of Cheerful Disposition.

We drove up the coast in the 1929 Whippet I had inherited from my uncle. Between the headlights of this green-colored coupe was a rod on which we could tie bags containing gopher skulls that we had soaked in water, drained, and wrapped in cheesecloth. As we drove along at thirty miles per hour, the bags of skulls swayed in the breeze and dried in a few hours.

At each site on our map we chose the most prosperous-looking farmhouse, knocked at the back door, and asked permission to set traps for gophers. Because Mary preferred to stay silent, it was I who became the butt of the farmers' jokes after I made this peculiar request.

"Tomorrow," I finally said to Mary, "it's *your* turn to talk."

Next morning we chose a large farmhouse in Loleta, a small town in Humboldt County. When the housewife answered our knock, Mary stood silent for what to me seemed an eternity and then blurted out, "We're from the museum."

"That's all right," the woman replied soothingly, fixing her eyes on Mary.

I butted in quickly. "We'd like permission to trap gophers on your farm."

Without moving a muscle, and still staring at Mary, the woman shouted, "Reuben, come here. Two girls here want some of our gophers!"

"What are you after?" Reuben asked, showing no surprise.

When we explained, he gave us permission to trap, provided we would give him a list the next day of the animals we had caught.

"We have moles as well as gophers here," he explained. "You'll never catch both animals in the same trap set, though, because a gopher is a rodent, but a mole is . . . is something entirely different."

Early next morning Mary found she had indeed caught a gopher and a mole in two sides of the same trap set. When she dangled the dead animals, still in the traps, before Reuben, his eyes widened but he said nothing.

We thanked him for his cooperation and were starting to walk away when he called after us, "The Woolgrowers are having a picnic this afternoon. Would you girls like to come?"

We thanked him but declined, and we continued north to an auto camp, where we rented a cabin for a dollar per night. After dissecting the gophers and preparing museum skins, we put the skulls to soak in an empty soup can, then ate a light supper and got ready for bed.

As I sat down on the hard bed, I glanced at the label on the can of soaking skulls, and I had to laugh when I saw that it read, "Heinz's Noodle Soup"!

17

Up Trapper Creek

JENNIFER BOVÉ

We had hiked the narrow dirt trail into Trapper Creek Wilderness for what must have been close to five miles, and it had rolled the whole way like an amusement park ride, one steep hill after another. According to Steven, this was supposed to be an easier way of getting to our remote study site than walking a mile and a half of stream channel, but Steven was an idiot. I only listened to him because, despite our identical ranks in the Forest Service hierarchy, he was the designated lead on that particular watershed sediment sampling project. Under any other circumstance I'm pretty sure I'd have spit in his eye.

I was loaded to the gills with eight five-gallon buckets, plastic lids, and a McNeil core sampler, which is a big steel cylinder in roughly the shape of a funnel. It's designed to collect cross-sectional substrate layers from stream bottoms, and we were using it to sample gravel in potential anadromous Steelhead spawning grounds along Trapper Creek.

There was no convenient way to carry the cumbersome sampler, so I'd rigged a shoulder strap onto it with my belt that was cutting into my neck with the persistent pressure of a dull blade. Even worse than the strap, however, were the blisters. I had planned to work *in* the creek that day, not to hike miles of uneven terrain, so I'd worn damp wading boots without socks. Every time I lifted a foot, moist leather skidded against the skin of my heel and ignited it in agony.

Ahead of me, Steven was carrying the same amount of equipment, and although his feet were padded with a pair of wool socks, he was a good bit smaller than me and looked as if he might just crumple like a paper doll under the bulk of his gear. You wouldn't catch him complaining about such things, though, no matter how much pain he was in. He kept his mouth shut. Trust me, it had nothing to do with stoicism; he just thought that making objections was "unprofessional." I, on the other hand, kept my discomforts to myself simply due to the fact that a fencepost would have been more inclined to empathize with me than Steven. I suppose that's why, as I suffered my swelling blisters in silence, I could think of nothing but how I wanted to hurl one of those buckets at him as if he were an oversized bowling pin. But I didn't. I just limped up one hill and down the next, cussing periodically under my breath.

The midmorning sun filtering down through the needles of the tall Doug firs was already too hot for it to be just the beginning of a bad day. Don't get me wrong—if Steven hadn't been with me, my outlook on life would have been very different. Trapper Creek Wilderness is like a 6,050-acre cathedral of conifers and sacred water in which one might pray that all places could yet be so pristine. Such preserves of riparian old-growth forest are becoming tragically more difficult to find in the Pacific Northwest. So if I'd been alone, I would have taken time for some personal veneration of that wild country. I would have gone at my own pace, stopping once in a while just to lean back and gaze upward or to smell sweet cottonwoods in the stream corridor below. But Steven wasn't into on-the-clock gratification of any kind. The only pleasure he seemed to allow himself was dragging me along in his miserable wake.

We descended an abrupt grade toward the creek, and I saw a shred of orange flagging on a tree limb just about the time I'd decided to declare mutiny.

I clambered ahead of Steven across the stream's rocky bank in anticipation of cooling my enraged heels in the water, and I had to bite my tongue to keep from yowling in pain when my boots slid awkwardly over the many melon-sized stones. Trapper wasn't an easy creek to traverse with the profusion of cobble and masses of large woody debris that signified the channel's health and stability. We were sampling it in part because of these attributes. Our goal was to obtain baseline data from Trapper to use as a comparison for other spawning habitat in more disturbed stream reaches. In spite of its many obstacles, though, I still would have chosen to walk the creek over Steven's five-mile trail.

I stood in the stream and waited for him to unpack his equipment. He was in charge of the data, so it was his job to record the date, site number, GPS coordinates, flow conditions, and number of samples we collected. When my feet reached an acceptable degree of numbness and Steven was ready with the GPS unit, I got to work, positioning my core sampler in the tail end of the pool.

When you're using a McNeil sampler, it's always a toss-up whether or not you'll get a sample on the first try. With this in mind, I stood downstream of the cylinder, held it on either side by the metal post handles, and jammed the end of it into the gravel. Then I started to twist it and push downward, forcing the small core cylinder deeper into the stream bottom. Everything proceeded as hoped until I hit a rock that wouldn't give. No matter how I tried to finagle it, that rock refused to cooperate by either moving into the cylinder or getting out of the way. I had no choice but to start all over again.

I moved a few feet upstream to avoid kicking excess silt into my new sample and began to dig, swivel, churn, and shove the sampler into the streambed again. The sound of the steel cylinder scraping against the gravel had become so familiar that I heard it in my sleep. Occasionally, I wouldn't get a good sample on the second try, or even the third, but that time I got lucky and was able to force the sampler through the rocks until the entire core cylinder was buried to the brim.

I began filling the first of my buckets with rocks from the sampler. It was easy enough to grab the big ones, but when I had to dig up handfuls of sand and gravel, it got pretty rough on my fingers. Gloves didn't help, even leather ones, because they wore through in no time. So I clawed my way down as gently as possible, transferring rocks into the bucket with care so as not to spill even the most minute sediment particles into the stream.

"Will you give me a hand?" I asked Steven.

He was still scribbling something on the data sheet—in all likelihood a crude sketch of my ass in the air. I can't imagine it was any coincidence that I caught him looking my way nearly every time I was bent over the sampler. He nodded and fumbled the clipboard into his backpack.

While Steven held the bucket, I slid the plunger component of the McNeil down into the sampler. It sealed all the water, sand, and silt from the sample inside the cylinder so I could pull it up and retain every last bit of sediment that was suspended in the water. The plunger sounded just like a toilet flushing when I shoved it into the sampler, prompting Steven to share a story about cleaning campus bathrooms to pay his way through college. But when he saw that he was not going to get much of a reaction out of me, he readied the bucket. I lurched the sampler from the streambed and dumped the remaining sediment-clouded water into the bucket with the rocks.

We meandered downstream throughout the day, repeating the sampling procedure until we had filled all sixteen buckets with substrate representative of Trapper Creek spawning habitat. That's what I did for a living. I carried empty buckets up streams, filled them with rocks, and lugged them back to the ranger station, where Steven would eventually perform particle size analyses.

I was busy thinking about the other stuff that people might be doing in the world for ten bucks an hour when I saw something that looked, at first glance, like a thick, rangy hound dog wading into the water not twenty yards upstream. Steven was finishing the labels on our last two buckets so he didn't take notice, and I found myself unable to speak. It took me only a second of wondering how a dog got all the way out here, miles from any semblance of civilization, before I realized the animal was not a dog at all—it was a cougar.

She stopped to look at me while two plump kits caught up with her. I watched the water rush past her forelegs, and it was evident, even from a distance, that one of her legs was as big as both of my calves.

"Hey," I finally whispered.

But Steven didn't look up.

Without taking my eyes off of the cat, I swatted the pen from his hand. "Steven, look. A cougar."

She turned toward us, lowered her head, and folded her ears flat in what I understood to be a warning stance. The kittens tumbled to a halt, attuned to their mother's signals.

"*Shit!*" Steven wailed, chucking gear all over the place in a sudden paroxysm of terror.

And then, before I could say another word, before we could formulate some kind of plan, he turned tail and ran. Just like that, he abandoned me there with the data, the buckets of rock, and of all things, a *mother* cougar. The only piece of equipment he managed to take with him, I quickly realized, was the radio. I didn't risk turning around to watch him go, but the rapidly dissipating sound of his boots clambering down the rocky channel told me he had no intention of waiting for me.

As I stood there alone, eye-to-eye with the animal, everything I'd ever heard about encountering a cougar flooded any thought of Steven from my mind. For one thing, I knew that a cougar was more likely to flee than to attack, but this girl showed no signs of backing down. Her stance was broad and powerful in defense of her young, and I was afraid that if I flinched, she might just tackle me in one easy leap. The thing you may not understand, unless you've ever been in that position, is how your survival instinct will erupt through your adrenals and shriek at you to escape. By standing still, I was fighting nature with every muscle in my body.

I yelled something at her, and though I don't think intelligible words came out, the kittens backed toward the woods a bit. Mama, however, held her ground.

Sweat was tickling my forehead and stinging the corners of my eyes. I couldn't help it—I started to panic.

I pick up the heaviest stone I could lift with one hand and hurled it at her. It splashed into the creek several yards shy of her legs, so I threw another, and another.

Only when, with cautious deliberation, she began to retreat did I dare pause to catch a breath. Trembling, I continued to resist motion, panting and waiting and standing my ground. The cougar visibly withdrew her gestures of warning. Her head raised, her ears came forward. I must have convinced her that I was too much trouble to deal with because she waded back to her kits and led them quietly off into the woods. I didn't move until several minutes after I watched the tip of her long tail disappear into the brush, and then I slowly dropped to my knees on the gravel bank.

It took a few minutes to shake off the encounter, but I had no other choice than to start moving. In spite of my suspicion that every snapping twig could have been the cougar coming back for me, I collected the gear that Steven had flung about the bank. It suddenly struck me as hilarious when I remembered the sight of his face contorted into that ridiculous wail, and I started giggling until my sides hurt and I had to sit down again.

Finally, I pulled Steven's pack onto my back and slung the core sampler over my shoulder, and then I looked at the buckets before me. We had left four buckets at each of our quarter-mile sampling intervals along the way, intending to tote only the four from this site back to the station, and then we'd collect the others the following week. But I could carry only two of them by myself. I squatted down and lifted the buckets, one in each hand, estimating their weight to be at least forty pounds each with all the gravel and water they contained. They were heavy, but because I was more interested in carrying them out of principle than obligation, I bit the bullet and began the slow, clumsy journey downriver.

It took me probably two full hours to go a mile and a half down the rugged Trapper Creek streambed. I tried to assemble a makeshift yoke with an alder branch to help me haul the buckets, but they wouldn't stay in place even when I tethered them with an entire roll of flagging, so I gave up the idea. There were some spots along the creek where I had to drag them (without spilling a drop) underneath fallen trees, and there were other places where I was forced to take to the woods to circumvent impassable logjams. But I kept at it, taking breaks whenever I needed to. I even stopped for a snack atop the mossy back of a giant downed cedar from which I could peer into a clear pool and watch fingerling Steelhead swimming below. It wasn't until I spied the late afternoon sunlight gleaming off the windows of our pistachio-green pickup that I decided to put on the afterburners. I've been known to get a second wind in the home stretch, even with aching shoulders and wrists pulled nearly from their sockets.

And there was Steven, sitting behind the wheel of the truck. He refused to acknowledge me. He just kept staring straight ahead.

I hefted the buckets and the McNeil sampler into the truck bed and hopped up onto the tailgate to pull off my wet boots. My feet were pale white and so waterlogged that my sandals were snug when I slipped them on, but the blisters had long since sloughed off and bled out, and the soft suede shoes welcomed my battered soles.

Just to watch Steven jump, I slammed the tailgate closed. He wouldn't even meet my eyes in the rearview mirror, though, so I gave up and climbed quietly into the passenger seat. I knew he wanted to argue with me about self-preservation and how I should have been smart enough to haul ass after him. What we both knew, though, sitting there in the stifling silence of the truck, was that he could have gotten us killed running off as he did, and that he would be going back to get the rest of the buckets by himself.

I rolled down the window, inhaling the scent of the cottonwoods deep into my lungs as I yawned and sank into the seat. It was, after all, just another day in the field.

18

Love in the Margin:
Finding Refuge in the Blue Mountains

JOSEPH L. EBERSOLE

Before I had even arrived in Wallowa County, Oregon, I knew the situation well enough to have told you why salmon were in trouble there and who was to blame. Resource-extracting industries had wrecked the stream habitats where salmon spawned and reared. Ranchers and loggers had altered nursery streams by eliminating the trees whose shade kept the water cool, draining water for crop irrigation, and eroding the banks so badly that the streams had all but washed away into shallow, sediment-filled ditches. These exploitive practices had driven several races of salmon to extinction and threatened the remaining few that were already weakened by overfishing and the gauntlet of Columbia and Snake River dams. It was ugly but simple: settlers had come and, within the span of a single human lifetime, had exterminated most of what was wild in the county.

On the up side, groups of people were mobilizing to restore the habitat of the Blue Mountains, to bring back extirpated wildlife and let the big trees grow again. The Wilderness Society, Hells Canyon Preservation Council, and other environmental groups were actively working to stop the destruction of forests, prairies, and watersheds by introducing bills into state and federal legislatures. If only these groups could overcome the vast inertia of an embedded resource-based industry, there would be hope for salmon.

From my perspective as a student of fisheries at Oregon State University, salmon were central. I was heading to Wallowa County to study the fish and provide the science that could help save them from those who were laying waste to the wilderness. I wanted to dive right into data collection, which I was certain would support my assumptions. My adviser, however, pulled me back.

"Slow down," he said. "Take some time to get to know the watershed. Muck around in the streams. See what kind of variation you can observe in the character of the landscape."

This was too vague for me; I wanted to know what to measure.

"Just get to know the place," he said, providing little help.

The following weeks of "mucking around" were, despite my anxiety, wonderful. I hiked knife-edged ridges, twisting canyons, and hidden valleys of lush, rolling prairie. I saw elk and deer everywhere in the late spring forests. It seemed a land of great promise.

During the weekdays of that summer I camped wherever I found myself. On weekends I came into the county seat to read in the library. To my delight, I discovered that the town was blessed with a thorough and prolific historian, and I read her books of stories of early Wallowa County. The librarian, impressed with an outsider's interest in local history, introduced herself. Next thing I knew, I was joining Cathy and her husband, Dave, for dinner. Over beer they pulled from me my reasons for being in the county. As it turned out, they were friends with a couple who managed a ranch smack dab in the middle of my study area, and they thought it would be a great idea for me to meet them. And so, by the time pie and coffee were being served, I was visiting with Mike and Judy, who insisted that we get together out at the ranch and take a look the creek. I nervously agreed, smiling good-naturedly. I couldn't turn down the offer, but I knew I wasn't ready to get into a conflict over salmon and cattle with a larger-than-life rancher. Mike asked me how I felt about the "whole cow/salmon thing," and I avoided what I had imagined to be a possible scene by answering that it certainly was a tragedy. Who could disagree?

In the weeks that ensued, we danced around the politics of salmon and cows and government regulation. Well, I danced; Mike was to the point. "So what makes you so sure that my cows are killing your salmon?" he asked me.

I was surprised that he was so sure I believed this, but I stuck to his question. Even though I tried to be truthful, my words came out sideways. I deferred to the dams, to complexity.

"It's no one thing, Mike. The Columbia's a changed river. Splash-dams have altered the creeks in these hills. We're in a drought." I added something noncommittal about grazing and riparian vegetation, saying, "It's possible that these slicked-off streams are warmer now than they used to be. That probably ain't good for salmon."

How quickly I found myself using his language and mannerisms! I talked slowly. I squinted at the cheatgrass-covered hills while we rested against the pickup hood. I nearly hankered for a chew. If I'd had a cowboy hat, I'm sure I would've tugged the brim when I leaned back to speak. But there was no denying that I was Joe Scientist. Unlike Mike, I qualified everything and was never absolute. I never said what I truly felt. I never told him how the shallow, silted stream channels outraged me; that I wanted to throw rocks at the blasted cows; that there were crimes committed here, and maybe not all of them were past history or someone else's violence.

Mike wasn't afraid to say what he felt. He swore at the environmentalists who had tried to close the Forest Service allotment that his cattle grazed. He made racist jokes. The best I could do at these times was smile at him, saying something ineffective like, "Aw, Mike, you know that's not the way it is."

I wanted to call his bluff, to curse back at him, but I couldn't.

At the same time, despite Mike's politics and manners, I was coming to enjoy his company. Mike knew the canyons like the back of his scarred hands. I continued to visit the ranch and learned a lot from him that summer. Still, I seethed inside while I walked streambanks denuded by cattle. In retrospect, I can see that by withholding my honest opinion from Mike, I merely contributed to his distrust of scientists. I can't imagine how I thought I was fooling him, playing the objective biologist. You see, Mike was a poker player—a good one, I eventually learned. He read me like the map of the canyons he didn't need.

As the summer progressed, I watched the water in the creeks drop, exposing sun-bleached rocks that hurt my eyes. What had been green either wilted or was eaten by the damn cows. I spent more time hiking into the high country on the weekends. The mountain streams were outside my study area, and I regretted that I had not insisted on doing my research up there. But my adviser had suggested I go where the problems were, to what are called "marginal habitats." These are the edges of a species' range, the limits of tolerance. For Wallowa County salmon, the margin meant the cow-trampled low-elevation canyons. But what was marginal for salmon was becoming intolerable for me. The population of juvenile salmon and Steelhead I had been observing through the early summer was rapidly disappearing. I could only hope that they were going upstream, finding cooler and deeper water somewhere out of the canyons. It was a logical wish for me to make. I, for one, wanted to get out of there.

Nancy Langston writes, in her book *Forest Dreams, Forest Nightmares*, that the first white settlers in the Blue Mountains "lost their hearts" when they encountered the loveliness of the land. These same hills are lovely still. But when they are seen closely and with a knowing eye, it is obvious that today something is terribly wrong. The big pines are mostly gone, springs and wet meadows have dried up, and salmon no longer teem in the rivers.

The early settlers may have lost their hearts, but they did not lose their heads. While proclaiming love for this new home, the settlers retained a will to transform what was already good into something that was "perfect." They wanted the best that the land could provide. They measured its possibilities in terms of human idealism, but their efforts to bring an abstraction of Eden to bear on a physical place of water, stone, and forest wrought havoc on the land. The settlers' hands, groping for paradise, inevitably strangled what they loved.

Today, after decades of fire suppression, heavy logging, and overgrazing, the Blue Mountains are something less than what they were a century ago. Yet those who were responsible for degrading this land allegedly loved it. *Love* is certainly not the first word to come to my mind when I consider human relationships to this region. I prefer to use words like *greed* and *ignorance*. Love as a potential motive for people's actions in the Blues is troubling to me. For if the settlers who began the cycles of overuse loved the land, and I claim to love the land, then what distinguishes us? Love for the land, I have assumed, is what separates people like myself from profiteers. Love for the land has been my assurance of righteousness and the guarantee that I could do no harm. Allowing that the others may too have loved this land challenges all my assumptions.

The salmon and trout I was monitoring in the canyons were seemingly dying by the hundreds as the temperatures climbed in July. Mike must have seen that I was having morale troubles. He noticed how much time I was spending up in the cool, forested slopes of the surrounding wilderness where cattle and cowboys rarely ventured, where I would live if I were a salmon.

"Cows git all your salmon?" he growled at me, smiling, as we drank lemonade in his kitchen.

I felt like slugging him, but I just sat there looking at his mason jar filled with rattlesnake rattles.

"Worst drought I've seen for quite a while… " This was Mike's version of consolation. "Them fish are smart, though."

I looked at him, puzzled.

"You ever look under my bridge?" he asked.

I hadn't. I didn't like snorkeling that close to a feedlot.

"Fish are holed up there in this hot weather," he explained.

I'd heard of fish congregating at cold seeps and tributaries to find refuge from warm waters. I didn't really expect it down in the canyon, though; it seemed much too dry and hot. I had to see this for myself.

I drove out to the creek and stopped on the bridge. As I leaned over the edge, my shadow fell across the stream bottom, and its movement spooked a dense school of fish. Within moments the school resettled into a close-packed bunch, focused over a pocket of clean gravel. I had never seen Steelhead and salmon school so closely together, and immediately I knew these were fish finding refuge. They were fish at the edge of their range, in the margin, hanging on under a rancher's bridge—all within pooping distance of his workhorses and milk cow.

My research after that day made a quick about-face. I started look-ing for seeps, and in the looking found them everywhere. Most were tiny, the size of a washbasin, but all were cold—occasionally even colder than the mountain streams. Some held only one fish, others had room for several. The seep under Mike's bridge would prove exceptional; it held 97 salmon and Steelhead one July afternoon. The fish had not left the canyon as I had, and as I thought they should have. They were in their own place, doing what they did best—getting by, no help from me. And I nearly missed it. The battered landscape, which had appeared at first glance to be dying, held life in small miraculous pockets.

How can I claim to love a land I so quickly dismissed? When the canyons withered, I fled to the mountains. I envisioned the canyon streams as mere shells of their former selves. In my mind I had dimin-ished them to something not worthy of my attention. Held against my own ideal of pristine greenery and thick runs of salmon, the canyon streams just didn't measure up, and I abandoned them. Yet they were not dead, not entirely ruined. Grossly mistreated, changed, stripped of vegetation, yes. But did this mean they did not deserve my love?

When I first came to the canyons, I thought I loved the land. But I mostly loved an abstract idea that had something to do with images of scenic beauty: wild rivers, rolling prairies, sharp granite peaks. My "love" knew little of the land as it most often is. How could it? I had not experienced the whole truth of the place; I had seen only distorted fragments. I saw death and ugliness in the overused canyons, and when the land failed to meet my expectations, I left it. Now I understand that the canyons remain alive, with secret graces and refuges that I've only begun to see.

I am going back to the Blue Mountains. Like the settlers who sought gain from the soil, fish, and grasses there, I go with my own set of great expectations. However, I can't help but wonder if researchers like myself, in our fervent attempts to save the land, might become just another obsessive force of control over it. Will this latest round of good intentions prove just as disastrous as the last? I don't know. Time may tell. In the meantime, you will find me out on Mike's ranch. He needs some fence mended, and he hasn't been able to get to it since his knee went out. And now, since he won't be able to chase me down and kill me, I'm going to tell him in livid detail how I really feel about the whole cow/salmon thing.

I owe it to him.

19

The Clinch River Diner

CHARLES F. SAYLOR

One Sunday, more than twenty years ago, several other biologists and I were surveying mussel populations on the Clinch River in northeast Tennessee. I can think of no good reason why we were working on a Sunday, but biological work never seems to be bound by days of the week. It was January, and we were snorkeling to locate mussels buried in the riverbed. The river ran cold and clear and may have even had a little ice around the edges. Swimming in near-freezing water requires some degree of insanity, but we were young, and we were wearing heavy-duty wet suits. In our minds, we were invincible. We sometimes even enjoyed playing like giant otters in the ice and the mud. But on this day the water seemed unusually cold. Overcast skies blocked the warmth of the sun, and a frigid wind chilled us to the bone whenever we came out of the water. After an hour in the water our hands and feet were numb. It was numbness beyond pain, numbness that made walking difficult because you weren't sure where your feet were, or if you even had feet. It was numbness that turned fingers and hands into useless clubs. We shivered uncontrollably and cursed the cold. I tried to raise the crew's morale by talking up the hot coffee and hot chocolate we would enjoy as soon as the work was finished. My companions, however, took this as a cruel joke. No one had thought to bring a thermos, and we all knew that there probably wasn't a store or restaurant within an hour's drive.

In between the shivering and cursing, my thoughts drifted back to earlier years when I'd worked this site. Less than a mile up the river there had been an old country store and a gas station. It was a cozy little store run by friendly folks who would have welcomed a bunch of cold biologists to come in and stand by the pot-bellied stove and sip some hot coffee and talk about the river and maybe even swap some jokes. It didn't seem so long ago that I was in there listening to a conversation between a young farm boy, the storekeeper, and a rather stout farm girl.

The boy asked the storekeeper, "Do ya know how to tell when ya see a level-headed woman?"

The storekeeper answered, "Don't know."

Barely holding off his laughter, the boy delivered the punch line, "Well, a level-headed woman has tobacco juice runnin' down both sides of her mouth."

Unfortunately for the boy, the girl beside him had tobacco juice running down only one side of her mouth. She was on him in a blink, swatting him across the back of the head with a newspaper as she chased him out the front door.

It didn't seem so long ago, but it was too long ago to help us. The store had closed years ago.

※ ※ ※

More shivering and more cursing, and my thoughts drifted back even further to the year I'd first visited this site. It was a hot summer's day, and working in the river was a cool relief. My crew and I had traveled to the site on a back road that ran along the far side of the river. I had not traveled that road since then, but I recalled a small diner by the road not far from the site. I could almost taste the hot hamburgers and French fries I had eaten there that day.

The memory was a little fuzzy as I stood on that cold riverbank, and I wasn't sure that it might not just be a dream born of wishful thinking. Anyway, I mustered up my optimism and started preaching to the other biologists about the wonderful little diner just down the river. Once again, my line of talk was not well received. There were some negative comments like, "I've been all up and down this river and I've never seen a diner," and "The cold has finally gotten to your brain and you're delirious." And finally there were challenges to put my money where my mouth was. Bets were made. Hot drinks and hot food were wagered.

If nothing else, all the talk took our minds off the cold until we could finish our work. When we did finally finish, most of us were so cold that we chose to wait to change into dry clothes until we found shelter from the wind, shelter like a diner restroom. In no time, we were driving down the back road on the far side of the river. Less than a mile down the road we came upon the diner. No one was more surprised than me. It was beautiful, an elongated white block building with large picture windows looking out over the river and a Coca-Cola sign hanging over the parking lot out front. And most encouraging, the lights were on, and there was a car parked out front, which told me the place was open for business. Still dripping a little mud and water, we herded ourselves through the front door. Inside, we found a man sitting at the counter drinking coffee and reading the morning paper. He appeared a little puzzled as he asked if he could help us.

I said, "Man, I need a cup of hot coffee."

"I'm sorry," he replied. "This is my home. The restaurant has been closed for several years now."

There was a distinct silence as the situation registered in our minds. Then we began apologizing. At the same time, the guys next to me threatened to choke me to death for getting them into such an embarrassing predicament. This was followed by bumping and shoving as we tried to make a graceful exit. If you can imagine having a pack of wet dogs storm into your house, you might have some idea of what this poor fellow was experiencing. Surprisingly, before we could leave, the man begged us to stay, saying that he had plenty of hot coffee and that if we wanted, he would get his wife to cook for us. We didn't accept his food offer, but we couldn't refuse the hot coffee. We sat and talked with this fellow for at least an hour. He had fallen on hard times. His restaurant had gone out of business after the furniture factory in Morristown had shut down and most of his customers, factory workers, had moved away. He was also having serious health problems. He didn't mention it, but it was apparent in his eyes that his world had become very lonely. Some say laughter is good medicine, though, so I hope he found some relief in our conversation.

As we said good-bye, I couldn't help but feel that everyone at the diner had just gotten warmed up in one way or another.

20

The science of Islands

SCOTT STOLLERY

These are the tools I use to fight off extinction:
a machete, a mist net, a caliper, and a syringe.
It hardly seems fair.

We enter the jungle at first light, thick
miasma coats our boots like finger paintings
done by a mischievous child.
Faint smells of broccoli and pig piss
hover in the air,
maybe orchids, wild lilies,
the end of the Hawaiian night.
It is painfully dense in here, ancient and wet:
I imagine that this jungle is like the mind of William Burroughs,
one idea growing on top of another
each seeking its own light,
a naked extension of reality, twisted
cut up
and pasted together
grown thick with rain and time.
A machete will work for today
but tomorrow the path will be gone.
Just the way the author wants it.

The sounds that rumble through the trees
are not the native passerines,
but jet planes.
They glide on Kona winds
like geese to a northeastern pond.
Only these geese don't care about the seasons
or know when to leave.
They keep coming and coming,
exploiting and breeding,
all little Kurtzes trying to swallow the world,
crapping out such great ideas as
sugarcane, mongoose, apple snails, mosquitoes, pigs, goats, pheasants,
*   ginger, rats, feral cats, pineapple, cars, donkeys, cattle, sheep,*
caged birds, guava, roads, and Wal-Mart.
And now a crew of ten people,
banding birds at a ten-by-ten-foot station
in the middle of the Pacific,
is supposed to fix all this.
This is the science of islands.

Each day at home,
where the banana trees and the morning glories
wage war for survival,
I sweep the kitchen floor to maintain my sanity.
There is something concrete about the neat little piles
of dirt freckling the white tile,
something manageable about dumping it all in the trash.
It is the one and only thing I can control, it seems.
And this is the science of islands.

Extinction is worse than death
for it holds no promise in its great black mouth.
It is a loneliness that cannot be seen with microscopes
or measured with scientific tools.
But each morning I hear the sweet call of the birds:
Apapane, brave Amakihi,
a blood-red gush of Iiwi, and the Omao.
Each one a miracle, fighting and singing,
lost in the art of life.
I will continue to enter the deepest jungle for them
and will listen closely to their stories
as we rise in elevation and breathe the bluish ethers,
until at last we find the heart of the islands.

21

Tigerland

ERIC DINERSTEIN

When I first learned that I was destined for Nepal as a Peace Corps volunteer, a place others associate with the mythical Shangri-La, I assumed I would be studying snow leopards or wild yaks in the vicinity of Mount Everest. I would be adopted by friendly Sherpas or study under Tibetan Buddhist monks, and maybe never return. However, when the Peace Corps information packet came, I realized that something was seriously wrong with my understanding of the offer. There were no pictures of mountain parks or information on snow leopards or monasteries in the clouds. Instead, there were crude drawings of tigers, rhinoceros, and elephants. My knowledge of Asian wildlife was pretty meager, but I knew there were no tigers or rhinos in the snowfields of the High Himalayas. As it turned out, I was destined for the Terai zone, Nepal's strip of lush (and unbearably hot) lowland habitat that sits at the base of the Himalayas.

Mel Sunquist, the quiet, serious tiger specialist we had met for the first time that afternoon, motioned for the Nepalese driver to stop. We were about to do something illegal, as all researchers were required to be out of the park by sundown. Quietly, we stepped out of the Jeep and onto the main track that runs through Royal Chitwan National Park. Mel's tigers, the first ever to be fitted with radiotelemetry collars, were on the prowl on this hot spring night in April 1975. My fellow volunteers were happy to tag along, and even more delighted to escape the crowded, dusty bazaars of Kathmandu and gain valuable jungle experience with a *tigerwallah* like Mel. Soon we were to be posted as survey ecologists in some of the most remote and untamed places in Asia. After two months of relentless language training—conjugating Nepali verbs and being tested on the local names for tiger (*baagh*), elephant (*haathi*), and Sloth Bear (*bhalu*)—it was high time to see a *baagh* up close. We stood silently in the receding heat of the evening, listening to the strange percussion of Long-tailed Nightjars and the distant alarm barks of Spotted Deer. I felt a million miles from home. Only a few months back we were faceless biology undergrads in the United States. Now here we were dressed in army surplus pants and T-shirts, a platoon of young American field biologists stationed in the heart of Tigerland.

Mel switched on the receiver, raised an antenna over his head, and began slowly rotating it in a wide arc, the ballet movement repeated by hundreds of field biologists before him as they tried to locate their study animals, except that he was trying to locate an example of the largest terrestrial carnivore on Earth. We felt no small amount of trepidation. Mel's Nepalese co-investigator, Kirti Man Tamang, was hospitalized back in Kathmandu, recovering from being pulled out of a tree and mauled by a tigress only a few weeks earlier. Following the conventional wisdom that tigers were reluctant tree climbers, Kirti was wedged into the crotch of a tree, waiting for the right moment to fire a tranquilizing dart into a tigress. The tigress decided that Kirti's close presence was a threat to her three young cubs, so she challenged conventional wisdom and scaled the tree. The thought of Kirti lying helpless on the ground, his legs badly shredded and an angry tigress standing over him, made me think twice about wandering far from the Jeep.

The tiger Mel was seeking was barely within range, so he tried the frequency of another. "I think we have a tigress very close by." His voice was animated, or at least as animated as possible given his Minnesota origins. Within seconds of Mel's warning, a fierce scuffle between the tigress and a large deer had us all scrambling back into the Jeep. The tigress in question had been right next to us, lying in ambush for a Sambar, or Indian forest deer (whose carcass Mel discovered the next day). Hidden from view by the green wall of Chitwan's elephant grass, this secretive predator was revealed only by her snarl and the piercing beeps echoing from Mel's headphones.

Back at the Smithsonian research camp, over glasses of warm Coke spiked with local rum, we chattered away about our first adventure with the King of the Terai Jungle. Yet the night wasn't over. We shifted from the eating area to a small clearing near the banks of the Rapti River, where we had pitched our tents on arrival that afternoon. American volunteers were paired up with Nepali language instructors as tent mates. All our instructors were well-educated, charming Kathmandu dwellers as new to life in the lowland jungles as we were. My favorite teacher was Narayan Kazi Shrestha, a bright, fun-loving man who eventually bestowed on me the gift of fluency in another language and the self-confidence that comes with it. But on this night my tent mate was Surya Sharma, a studious high-caste Brahmin in his early twenties and the son of a famous Nepalese judge. As we drifted off to sleep, the sound of loud chewing and lip smacking stirred us awake. Surya peered out of the insect netting and experienced a mild shock—grazing next to us was an enormous Greater One-horned Rhinoceros accompanied by a calf. Surya reached over and clutched my arm, "Rhinos!" he whispered fearfully, using the English rather than the Nepali word (*gaida*), perhaps figuring that in light of the danger at hand, this was no time for a vocabulary drill. We had been warned earlier that rhinos routinely trample and kill several tourists each year. I peeked through the fly mesh and put the index finger of my free hand to my lips. Surya's grip tightened. After a while the mother and calf wandered off, but the interlopers left a lasting impression on both of us. For me, it was my first face-to-face experience with a creature I would eventually devote years of my life to studying and conserving. Surya seemed quite shaken by our little jamboree, and the effect was equally overwhelming: he never taught a Peace Corps language training again and, when our program was over, went straight into law school.

When our training period ended, I headed for the Royal Karnali-Bardia Wildlife Reserve. I had been handed an introductory letter to the Bardia park warden that detailed my mission, a single sentence hastily typed by Nepal's senior ecologist, Dr. Hemanta Mishra: *You are to census the tiger population in Bardia and to conduct other wildlife inventories as appropriate* (translation: *Get out of my hair and see if you can do something useful out there*).

Today Bardia is easily accessible by an excellent all-weather road, but back then it was Nepal's version of the outback. A posting there was considered by Nepalese officials to be banishment rather than a plum assignment. Bardia lies one hundred eighty miles to the west of Chitwan and is one of the most spectacular wildlife reserves anywhere. Its western boundary is the Karnali, the wildest river that flows out of the Himalayas. The Karnali is filled with crocodiles, Gangetic Dolphins, and floodplain islands of rosewood and acacia. Wild tigers and their prey swim back and forth among these islands. In 1975 Bardia had never been properly surveyed and seemed like a destination that would fulfill my earliest childhood fantasies and my more recent academic training as a wildlife biologist.

The standard joke among former volunteers is that in the old days the Peace Corps parachuted you into the bush with little more than a Swiss army knife and a copy of JFK's *Profiles in Courage*. But being dropped from an airplane into Bardia wasn't an option. With the monsoon fast approaching, all the nearby grass airstrips were no longer serviceable. We had no alternative but to reach Bardia via four-wheel-drive Jeep. Together with Will Weber, my Peace Corps director, and Cliff Rice, a fellow Peace Corps volunteer slated to become my western neighbor as the ecologist of the Royal Suklaphanta Wildlife Reserve, we headed south from Kathmandu. There was no road across lowland Nepal in 1975. After crossing into India at Bhairawa, we would have to continue westward on the Indian side of the frontier. On our way from Kathmandu to the lowlands we found our route blocked by a massive landslide. Such natural disasters are a common feature in the Himalayas during the monsoon. Impatient Westerners become unglued by landslides because delays are often measured in days rather than hours. The Nepalese seem wonderfully preadapted to cope with the inconvenience, blessed as they are with a unique brand of happy-go-lucky fatalism. "*Ke garne?*" ("What to do?"), a popular phrase uttered with a smile and accompanied by a wave of the hand, could easily qualify as the national expression.

When we reached the landslide, people had been waiting for almost two days for the road to reopen, and a carnival-like atmosphere had already developed. Tall Westerners are conspicuous in a land where most people are very short. We quickly noticed another one of our kind sitting forlornly in a tea stall that was catering to the temporarily detained. It was Dietrich Schaaf, a Ph.D. student from Michigan State who had been in Suklaphanta for a year studying Swamp Deer, a highly endangered variety of the world's thirty-seven species of deer and the most renowned species of that reserve. Encountering a few more West-erners did little to cheer up Dietrich; he had been blocked for more than a day from reaching Kathmandu. He also seemed displeased to see Cliff, who was the only volunteer among us who held a master's degree and could claim a strong background in ungulate biology and behavior. In truth, he was the only one of us who was really academically prepared to take on his assignment. Unfortunately, what Cliff really wanted to do was to study Swamp Deer, and Dietrich had already staked out this species as his alone, making sure in a previous meeting that Cliff under-stood the arrangement. This was to be my first and far from last lesson in territoriality among biologists.

Eventually, the last boulder was pushed down the mountainside into the Trisuli River, and the bulldozer cleared a narrow route through the rubble; we were on our way to India. We raced the onset of the monsoon across Uttar Pradesh, driving as fast as we could while swerv-ing around imperturbable cows, horse-drawn buggies, ox carts, and countless people on bicycles.

Once we crossed back into Nepal below the town of Nepalganj, we met with the conservator of forests, who traced our route west to Bardia. There was only a dirt track, he said, and the rains had started already. I tuned out the rest of his warnings, pretending not to com-prehend what I didn't want to hear. Several hours later, after winching ourselves out of several streams, we gave up and headed back to Nepal-ganj and the conservator's office. I would have to wait for a Forest Department elephant to carry me and my belongings to Bardia.

Will and Cliff left me in order to reach Suklaphanta by nightfall. I walked over to a large building that turned out to be a USAID guest-house. It was currently empty except for a Nepalese caretaker, an ex-Gurkha soldier. I asked if I could stay there, and he replied that he couldn't let me without a letter. Anything that you wanted in Nepal, down to the smallest request, required an official letter, a *chitti*. And when you handed the *chitti* to a government officer, even if it boasted enough stamps of authenticity to guarantee your passage to Nirvana, he would hold it away from his face in feigned puzzlement, as if your request were preposterous or were written in Cyrillic or Chinese rather than in Devanagri, the script shared by Nepali and Hindi.

Besides the problem of not having a *chitti*, the ex-Gurkha said that the cost for a room at the guesthouse was about fifty dollars per night, almost half my monthly stipend. Now what? I thanked him and turned to leave when he asked me, in an innocent voice, "Do you know geometry?" He went and fetched a hardbound primer, a relic from the British Army education program. If I would teach him all about isosceles and right triangles and other such objects, he would let me stay for free while I waited for my elephant. I silently thanked Mrs. Luella Beebe, my tenth-grade geometry teacher, for sparing me a night of camping out with cobras and banded kraits in a patch of scrub jungle at the edge of town.

Two days later my elephant finally arrived. It was an old female, accompanied by her two sullen drivers. It takes several days to travel the fifty miles to Bardia, and this was a slow elephant. We had to set off immediately. I strapped my small shipping trunks on top of the elephant's saddle and took a seat between them. No sooner had we gotten under way than the rains began again. The elephant had trouble moving quickly through the mud that masqueraded as the seasonal dirt track between Nepalganj and Bardia. The second day also brought a punishing rain, and this time we halted our trek. Fortunately, we had reached a forest development project camp, and I was kindly hosted by a Danish and Nepalese forester team while we waited for the end of the deluge. The elephant drivers were eager to go on, but the Nepalese forester said that the past two days of downpours would make the next stretch difficult and that we would be prevented from crossing the Babai River, which separated us from the Bardia Reserve and its headquarters, Thakurdwara.

I had come prepared for such circumstances; in my backpack was a copy of *Huckleberry Finn*, and I was rereading the section where Huck and Jim were floating down the Mississippi, escaping from "civilized" society and slavery. I tried to picture my elephant ride as embarking on a Mark Twain journey to a land of adventure. But unlike Huck and Jim, I had a close destination and a research program to initiate. The thought of being stuck on this side of the Babai River, looking across at Bardia

Reserve until the end of the monsoon, made me nervous. What was the alternative—return to Kathmandu and wait until the rains stopped in September? Whatever happened to me over the next two years, I told myself, I had the option of treating each new challenge as a source of worry and defeat or as a new adventure. No matter what life in Nepal's jungles held for me, I vowed that I would always choose the latter.

When the rains stopped, I had a chance to test my new dedication to this philosophy of adventure. We saddled up our *haathi* and headed west. Within a few hours we reached the banks of the Babai, and to my dismay, the river was a deep brown torrent. Across the surging water beckoned the rosewood and acacia forests of Bardia. The drivers were determined to cross without delay. The mahout, sitting behind the elephant's head, urged her down the riverbank. She stalled at the water's edge, perhaps gauging the speed of the current or the stupidity of the humans sitting on her back. Watching the river rush by was dizzying. The mahout would have none of it. Whacking her with his stick across her broad forehead and muttering curses, he drove her forward. Within a second the elephant was up to her knees and elbows, then shoulders, and before I could tell the driver that we might want to reconsider our plan of attack, we were swept away. For a brief moment, all that was above water was the tip of the elephant's trunk and my head. Elephants are rather buoyant, however, and powerful swimmers. Within seconds the drivers, who held on to the saddle ropes, had us back on the riverbank. I immediately opened my steamer trunks to assess the damage. Fortunately, all my clothes, books, and camera gear were packed in plastic bags or inside waterproof biscuit tins and Tupperware, and everything but the shorts I was wearing had stayed dry.

I looked up the riverbank and noticed a small, open-sided thatch hut where a group of men were clustered. I walked up to them and saw that they were playing cards and drinking tea. I asked them in Nepali if they had seen what had happened. Somewhat startled that a Westerner had addressed them in their own tongue, they all smiled and offered me a seat and a cup of tea. "You tried to cross in the wrong place," they laughed. "Just wait a few hours. The Babai River is not like the mighty Karnali. It only drains the low mountains behind us. So when the sun shines, the river level will drop and you will cross the

Babai in this place in front of where we sit. *Aram garnos* [another national expression, which roughly translates "chill out"]."

Over tea the villagers' questions unfurled in a standard sequence I would soon grow accustomed to: "Are you American? How old are you? Are you married? What is your salary? How much land do you farm? How many cattle and buffalo do you own?" The villagers had no problem with my bachelorhood, even at the advanced age of twenty-two. But they were dumbfounded that I didn't grow my own rice back home or that my family kept no water buffalo or goats in their Miami condominium.

Eight hours after our first misadventure we arrived at reserve headquarters, a cluster of two whitewashed brick buildings and several thatch-roofed huts. I was met by Krishna Man Shrestha, the newly appointed park warden. The drivers removed my steamer trunks and, before I could even wave goodbye, turned the elephant back toward Nepalganj. I handed the warden my letter of introduction from Dr. Mishra, thankfully kept dry in a Ziploc bag. The warden put on his reading glasses and kept glancing back and forth from the letter to my face. He never altered his expression—a distinct frown made all the more severe by the scars of a childhood battle with smallpox. I told him in respectful Nepali that I was assigned to be the new survey biologist for the reserve and looked forward to working with him over the next two years.

Decades of service in the government bureaucracy had honed Krishna Man's skill at establishing hierarchy. The caste system was about to be applied, even to an extra-terrestrial like me. He pointed at a tiny two-room thatched hut on stilts and declared that I could share it with an army sergeant who was temporarily stationed in Bardia. He asked what was I going to do for food and how I would I eat. Had I brought any rations with me? I hadn't even thought that far ahead. I was also quite hungry, having consumed nothing all day except the tea and biscuits offered to me by the Nepalese villagers who witnessed our first attempt at the river crossing. The warden told me that I could share meals with him for a few days but then I would need to find a cook and figure things out on my own. The nearest bazaar was on the Indian border more than ten miles away.

I unloaded my few belongings onto the two shelves in my cubicle. The sergeant was off on holiday, but there were plenty of other occupants. At dusk the thatching emitted a cacophony of chirps and squeals. By nightfall it was alive with rats and a nursing colony of Indian Pipistrelles. Five years later I became enamored with bats and even studied them for my Ph.D. But my first experience with them was nothing short of traumatic. I nearly panicked when the juvenile Pipistrelles dropped on the floor or my mosquito net while learning to fly, or flitted through the narrow room, tight furry fists glued to leathery wings. The rats held their own in this hut of horrors, chewing on everything I absentmindedly failed to stow away each night in my steamer trunks. My mosquito netting doubled as a barrier to keep nocturnal rats and bats from scrambling over me like a piece of furniture. By candlelight I wrote in my journal the sentiment expressed by many other young volunteers, the sentence that attempts to dispel a premature wave of adult despair with the naive optimism of youth: *Just make it through the first month, and the rest will be easy.*

On the third morning a tall, handsome Nepalese man clambered up the stairs of my hut to marvel at the new carnival freak from America carried in by elephant. His name was Gagan Singh. In what seemed like one breath he told me that he was a game scout for the reserve and not to worry, that he would find me a cook and together he and the cook would go to the bazaar on the Indian border, buy all the pots and pans I needed to set up a kitchen and a month's worth of food rations (rice, lentils, potatoes, onions, sugar, salt, and tea), carry it all back to Thakurdwara, and when he returned he would help me get settled, guide me around the reserve, and show me where the tigers lived. To the rest of the world Gagan may have been a dark-skinned, low-caste Sunwar (the low-ranking goldsmith caste) who had not studied beyond the third grade. But to an unnerved twenty-two-year-old foreigner, he was my Gunga Din.

Gagan returned late the next day as promised, accompanied by a young Tamang named Prem Bahadur, both of them carrying on their backs enormous loads of food and kitchen goods. Prem would be my cook, and over the next two years Gagan became my everything else: tiger tracker, jungle tutor, Tharu anthropologist and language instructor, older brother, and best friend. Gagan had grown up in the forest and was a true jungle savant who was more than willing to share his knowledge with me.

By the end of my first month I felt settled into a routine of fieldwork and exploration of the reserve. But things had taken a turn for the worse in my temporary hovel: besides the rats and bats, the roof leaked like a colander during every monsoon downpour. Whether it was the result of a shoddy thatch job or the diabolical shifting and swarming of the commensal vermin, I wanted out. So I made plans to build a native-style house at the edge of the jungle once the monsoon ended in October. Gagan helped me select the best site close to the banks of the Khoraha River, about five hundred yards from where I was currently residing. I envisioned a veranda where I would watch langur monkeys leap through the fig trees on the other side of the Khoraha and smooth Indian otters glissading down the riverbank. A leopard or jungle cat in my front yard would be perfect. It was time for some privacy, after having endured the scrutiny of my neighbors, who took note of my every move as if I were the object of biological research. My thatch-roofed palace would be ready by the rice harvest.

In July 1975 Gagan and I began a walking tour of Bardia. Some Westerners I met had hunted Sambar Deer along the edge of the reserve a decade earlier, but ecologically speaking, Bardia was a blank spot on a map of the world. Tigers were still present, but how many and how serious the effects of poaching had been remained unknown. Rhinoceros, on the other hand, had disappeared from Bardia's floodplain at least a century earlier. Gagan and I decided to begin by exploring the northern edge of the park at Chisapani and then head east to the Babai River, the site of my accidental baptism on the ride into Bardia.

After walking along the dirt tracks crisscrossing the reserve, I decided to use this superb road network to carry out my tiger census. I knew from everything I had read and all that I had seen so far that one rarely encountered tigers out in the open. I would have to employ an indirect method to estimate their numbers. Old junglewallahs know that tigers prefer to move along dirt roads and trails to patrol and mark the boundaries of their territories. Indian foresters and wildlife officials have always used the footprints of tigers, known as pugmarks, to count them. Some Indian wildlife officials claim that such counts are quite accurate if trained observers conduct the census. According to proponents of the pugmark technique, individual tigers can be distinguished by their size or by deformities in the pads of the feet.

After measuring about a dozen sets of pugmarks, I began to have my doubts about the validity of this approach. I observed, as did others, how the size and shape of the footprint changed markedly depending on the substrate. If a tiger stepped in mud or soft sand, the toes and pad spread out. If a tiger stepped in fine dust, the perfect substrate for taking measurements, there was less variation. Unfortunately, I found few perfect impressions in the ooze of the monsoon; the pugmark study was a washout. Furthermore, lying in the dirt to trace tiger tracks and then spending all evening calculating analyses of variance was dreadfully boring and a far cry from how I pictured spending my days in Bardia.

There had to be a way to census tigers other than using pugmarks, so I went back to the wildlife literature. I consulted my bible, *The Deer and the Tiger*, George Schaller's classic 1964 research account of the wildlife of Kanha National Park, India. Schaller has inspired more field biologists on more continents than any other living scientist. His love of natural history, his ability to endure hardship in the field, and his gift for quickly capturing the essence of a species' biological niche are world-renowned. By the end of my first six months in Bardia I must have read *The Deer and the Tiger* several times. Unfortunately, the quality of Schaller's scholarship and the clarity of his writing provided little resistance to the indiscriminate mildew of the Terai monsoon.

The book forced me to change my thinking about the importance of studying tigers versus the biology of their prey. Because tigers are such magnificent animals, most biologists are afflicted with serious research myopia. A disproportionate amount of effort is spent documenting the ecology of the glamorous carnivores while the more pedestrian prey are neglected. Work by Mel Sunquist on the nutritional requirements of female tigers revealed that it takes about two Barking Deer per week to fuel a breeding female tiger. If prey intake falls below this threshold, tigresses stop having cubs. Thus we needed more information on the biology of the tiger's prey, and especially the prey species' population dynamics. Schaller's book provided a model of how to go about studying not only Barking Deer but Spotted Deer, Hog Deer, Sambar, Swamp Deer, wild boar, and other species that find themselves on the tiger's menu.

Study of prey is critical for carnivore conservation. Scientists learned relatively early the deceptively simple ingredients of tiger management: set aside a sufficient area of suitable habitat with enough surface water and stop the poaching of native ungulates and tigers. Do this and tigers will come roaring back from the brink. Tigers and wolves are the two large predators that actually breed faster than their prey. It's also hard to drive tigers to extinction because they are resilient and wily. The best way to wipe them out is to eliminate the species on which they feed.

I tried several approaches to studying the tiger's prey. The most common method is to estimate the density of deer species by systematically counting droppings left behind on a randomly located series of study plots. There is an entire literature devoted to what a friend of mine calls turd biology, or turdology for short. A turdologist in Nepal must be able to distinguish the discrete clusters of small round droppings of Hog Deer, the more elongate pellets of Spotted Deer, the wider cylinders of Sambar, the even larger offerings of Swamp Deer, the vitamin-shaped tablets of hares, and the bonbons of wild boar. Then there are the droppings of tiger prey that return to the same sites to defecate (called latrines): the minute, comma-shaped pellets of Barking Deer and the Junior Mint–like clots left by Nilgai Antelope, the largest antelope of Asia. Our routine was as follows. First, Gagan and I cleared the plots of any droppings we found. Second, we revisited these plots at regular intervals to count and remove the pellet groups left behind by obliging ungulates that had no idea that with each defecation, they were contributing to our understanding of predator-prey dynamics.

If measuring tiger tracks was astoundingly dull, crawling on hands and knees to pick up pellets was a close second. Having extra workers helped us clear plots more quickly and allowed us to survey a larger number of sample areas. Sometimes Gagan would recruit Tharu friends by teasing them that we were going to pick up gold nuggets off the jungle floor.

The pellet counts yielded useful results on population densities of tiger prey. But the quality of the data didn't change the tiresome nature of this activity. Far more appealing were afternoon safaris spent counting tiger prey from the back of a Toyota pickup. From March through May, after the fires have burned through the grasslands, the Terai jungle resembles the savannas of northern Tanzania. During the hot season we would drive slowly in the late afternoon to count the herds of deer and boar that came out to graze on grass shoots or feed on fallen fruits. The most remarkable discovery was the rapid response of the deer population to improved conservation measures. The Nepalese Army assumed responsibility for the protection of the reserve in 1976, and its presence changed the behavior of both the wildlife and the poachers virtually overnight. I commonly heard gunshots when I first arrived in Bardia, and the easily spooked wildlife rarely ventured out in the daylight. Immediately after the army began patrolling, the gunshots ceased, and by the next year the wildlife moved about without fear. By the second year of my study we had trouble counting the vast herds of Spotted Deer that formed during the late afternoon grazing hours, and I was delighted to rediscover a small herd of Swamp Deer that had survived and came out of hiding, like refugees from a root cellar after a tragic episode of ethnic cleansing. As the poachers retreated, the nilgai, or blue bulls, were also staging a comeback. Conservation in South Asia does not require the generations of patience essential to witness the recovery of endangered rain-forest species or the fragile arctic tundra. Because of the rapid healing powers of the annual southwest monsoon, this is one landscape where nature quickly covers up human transgressions.

Another exciting means to observe wildlife was by floating the Karnali River from the gorge at Chisapani back to a landing near the park headquarters. Along the way one could stop and survey lovely islands covered entirely by rosewood trees, acacia, or a mixture of the two. These are the first tree species to colonize the floodplain because they fix nitrogen (both are legumes) and are able to grow with their roots partly submerged in water during the monsoon period. They also provide excellent habitat for deer, thereby attracting tigers.

Bardia is a prime example of one of the few remaining areas in Asia where tigers still occupy vast floodplains and river deltas. The only other outpost—the Sundarbans—has attracted an international reputation because the tigers there have become man-eaters. In that vast honeycomb of mangrove swamps and river delta on the border between Bangladesh and India, tigers routinely swim between islands during high tide in search of food. All too frequently this may include unfortunate villagers. There are many theories to explain the prevalence of this frightening behavior. One preposterous suggestion is that tigers have developed a taste for salty human flesh because they drink seawater. Another doubtful hypothesis is that Sundarbans tigers are more aggressive than tigers in other parts of India and Nepal. The simplest explanation is numeric: the relative scarcity of natural prey in mangrove swamps is supplemented by the high density of humans who invade daily to fish, collect honey, or cut wood for charcoal. The Sundarbans may have the highest ratio of human density to native prey density in any area of the tiger's range.

Man-eaters were nonexistent during my Peace Corps days in Bardia, and the Karnali floodplain was the best place to see a tiger. Our first float trip down the Karnali was made possible by the donation of a Zodiac raft from the United Nations Food and Agriculture Organization (FAO) project attached to the Department of National Parks and Wildlife. The head of the project, Frank Poppleton, was a graduate of the old school of East African game wardens. Frank was a commanding presence: well over six feet tall with a shock of white hair, brilliant blue eyes, khaki shorts, pale green kneesocks, and desert boots. Unfortunately for Frank and other white colonials residing in East Africa, their days were numbered with the coming of *uhuru* (African liberation). So their last refuge became the FAO and secondments to wildlife programs in the developing countries of Asia.

By early morning Krishna Man, Poppleton, Gagan, another game scout, and I hopped in the Zodiac and set off from Chisapani. After we had settled in the raft, Frank recounted an earlier raft trip he had taken down the Rapti River in Chitwan. Then he told me how he valued the work I was doing and how lucky I was to be out here in such a beautiful setting. The gentle pull of the river lulled me into reflections about the extent of my good fortune to spend two years in the middle of nowhere.

One of the goals of our river expedition was to locate gharial. Gharial are perhaps the most ancient of living crocodilians; their roots go so far back, they could have played pinochle with the dinosaurs. If crocodiles hold title as the most hideous-looking creatures on earth, then gharial are the Quasimodos among crocodilians. Their long snouts filled with about fifty pairs of sharp teeth, their beady eyes, and the giant knobs at the end of the noses of adult males are unlikely to inspire plush toy designers. However, if you are lucky enough to see one raising its prehistoric snout out of the deep brown murk of a monsoon river, then you are the rare naturalist who has experienced a firsthand glimpse into the Jurassic period.

At the time of our float trip, there were probably no more than two hundred gharial left in the wild, and Chitwan and Bardia were two of the last strongholds. Unlike the more terrestrially adept Nile crocodiles or American alligators, gharial are hopeless on land. Their weak forelegs make it hard for them to climb more than a few feet up the riverbanks to deposit their eggs, far below the more private nest sites beneath the tangle of shrubs at the edge of the jungle. In protein-starved South Asia, gharial eggs were not only easy to collect but also sought after by poor villagers, to eat themselves or sell to brokers. And if the villagers didn't rob the exposed nests, an unexpected pre-monsoon flood could wash away the eggs and that year's reproductive effort. For the past fifteen years, scientists had been collecting the eggs of gharial, hatching them in a captive breeding facility inside Royal Chitwan National Park, and then releasing the juveniles back into the rivers when they had grown large enough to fend off adjutant storks and other predators. Even then their future was far from secure. They had to locate a river stretch with enough fish to survive and breed, and this was becoming increasingly difficult in Nepal and India, given the pressure local fishermen exerted on all the major rivers. Places like Bardia that were strictly protected were the last hope for avoiding the total extinction of this ancient reptile.

On a sandbar in the middle of the river, Gagan's sharp eyes located a gharial. And then another. It was too early in the year to observe nesting behavior, but their presence so near Chisapani gave us hope. Today gharial have made an encouraging comeback and have been reintroduced in other rivers within their former range, thanks largely to the dedicated efforts of long-time gharial researchers Dr. Tirtha Man Maskey, now director-general of Nepal's national parks, and Ram Pritt Yadav, the senior warden of Chitwan for most of the past two decades. Their efforts illustrate what a few dedicated individuals can do to save a species from extinction.

Along the riverbank Frank spotted a pratincole and a wallcreeper, birds he was familiar with from East Africa. After a short stretch the vast floodplain of the Karnali opened before us. What began as a tame boat excursion had just crossed the threshold into adventure. Something surfaced ahead of the boat that Frank would never have observed on a wild African river. It then reappeared fifty feet away. We quickly spotted another creature not far from the first. I only glimpsed a long snout attached to a gray lump, but the sound it made as it broke the surface required no further confirmation. We had come upon a pair of Gangetic Dolphins. There are only six species of freshwater dolphins in the world. The Amazon and the Orinoco have their two species, one as pink as a flamingo. But freshwater dolphins reach their maximum diversity in Asia with four species, each placed in its own genus. The Indus, the Ganges, the Irawaddy, and the Yangtze Rivers all have their own distinct species.

We tried to paddle closer to the dolphins, but they kept their distance. Unlike the gharial, the Gangetic Dolphin has continued to decline over the last two decades, after the Indian government built barrages and dams restricting their movements. It's a tragedy that as of 2001, the population has dwindled to a few stalwart dolphins occupying only the Karnali in the west and the Kosi in the eastern part of Nepal. Gangetic Dolphins and the other freshwater dolphin species lack the charismatic personalities and the legions of Western admirers that rush to defend the cause of their marine relatives. Physically, the Gangetic Dolphin bears only the slightest resemblance to a spinner or bottlenose. Asian freshwater dolphins are practically blind, relying on echolocation to find their way and to catch fish in the muddy rivers of the monsoon. I used to walk the two miles from Thakurdwara to the Karnali during the warmest days of May to cool off in the water and to try to approach the small population of dolphins that hung around the area. I never succeeded in doing so, but the sight of them rolling through the current is etched permanently in my collection of wildlife memories from Bardia. Gangetic Dolphins are the first species I encountered in the wild that will probably go extinct in my lifetime.

Gagan had a strange expression on his face. Something was not right with him. Normally the toughest of junglewallahs, he seemed to grow uncomfortable by the minute, but being of such low caste he was too embarrassed to ask to stop the boat. Finally, his good sense and physical discomfort overcame his feeling of social inferiority, and he politely asked if we could stop at the next piece of land so he could relieve himself. We soon reached the first of a score of floodplain islands with a dense tree cover of rosewood and acacia. Gagan leaped out of the boat and headed for the nearest thicket. The rest of us followed suit. Gagan had disappeared for no more than three minutes when he came charging back, pulling up his shorts as he ran toward us. While searching for a quiet spot, he had practically stepped on a sleeping tiger.

We followed Gagan single file to the site of his encounter and quickly climbed up some rosewood trees to gain a better view. I looked to see that everyone was at least fifteen feet above the ground, the memory of Kirti Tamang's near-fatal accident still fresh in my mind. After a few minutes we grew bored with watching the large male stretched motionless in the grass below us. I thought it seemed too risky to have it wake up while we descended from the branches, so we decided that the safest course was to rouse the tiger. While shouting "Get up, King of the Jungle!" we clapped our hands loudly. The tiger must have been awake throughout the whole episode and just biding its time to make a run for it. With a great roar it exploded from a prone position and leaped what seemed like fifteen feet in one bound, fortunately in the opposite direction of where we perched. We scampered back to the raft and the safety of the river. After nearly eight months in Bardia I had finally seen my first tiger!

I was to see other tigers without the help of Gagan's indigestion to initiate contact. One of the techniques Schaller used in his study was to observe the behavior of ungulates and their predators from tree platforms, known in Nepali as *machans*. Gagan loved to construct things, and when I told him of my plan to erect a series of *machans* in each of the habitats in my study area, he willingly agreed to design and build them. He recruited two Tharu friends for this task, as well as my new cook, Kanchha. Within a short time we had erected thirteen *machans* scattered within four miles of my house. They were simple structures: the floor was fashioned from a few wood planks lashed to tree branches by vines and rope; the roof was the starlit night sky. I was the Peace Corps' first semi-arboreal volunteer.

My life quickly settled into a new routine: a quick dinner at three-thirty in the afternoon in order to be out walking and reach a tree platform within the hour. On the way to my favorite *machan*, I would pass by the prized perch of a jungle owlet in a grove of acacia. I sometimes nodded to this bold little fellow, a night watchman of Bardia, as I headed out to assume the late shift. In the spring I would occasionally stop to witness the wild mating dance of the peacocks. One of the most wary of local residents, peacocks roost at night on the outermost branches of the kapok trees where even the agile leopards can't stalk them. But when a male peafowl performs his spring étude in a jungle clearing, he lapses into a choreographed frenzy of vibrating feathers, an easy target for a hungry predator. I have picked up the discarded costume of tail feathers often enough in the spring to know the mortal risks of wild courtship dancing.

When I reached my *machan*, I would take off my backpack and remove a length of rope, tying one end to a belt loop in my shorts and the other to the pack frame. After climbing up to the *machan*, I hauled the backpack up after me. Then I would quickly arrange my foam pad and sleeping bag, drink some water, and wait until the sun started to sink and the deer and boar emerged from the shadows. Around five o'clock the first herds of Spotted Deer began filtering into the grasslands from the forest edge. Spotted Deer, also known as Axis Deer or Chital, are the most abundant deer in the subcontinent and certainly rank among the most beautiful of all cervids. The fawns of most deer species are spotted at birth, but this species and Fallow Deer are the only two that keep their spots into adulthood. Spotted Deer males have enormous antlers, reportedly the largest in relation to their body size among the thirty-seven species of deer. They are also the favorite prey for Bardia's tigers.

During more than one hundred overnights in *machans*, I took detailed observations on Spotted Deer, watching their breeding behavior and noting what they ate as they grazed and plucked leaves below me at dawn and before dusk. To help me interpret what I saw, I overcame my initial shyness and struck up correspondences with John Eisenberg, an encyclopedia of mammalogy, and Richard Taber, one of the world's leading deer biologists, who was to become my graduate adviser at the University of Washington. I was still too intimidated to write to Schaller. Based on my observations and data from the exhausting pellet count studies, I published a series of papers on the ecology of tiger prey and their habitats in Bardia when I returned to graduate school. But the scientific language of the papers failed to capture a powerful emotion that grew stronger with each passing day: I was witnessing the recovery of a national park. The deer had increased in large numbers, and the tigers prowled without fear of humans.

I have many vivid memories of my nights in the tree platforms. Like the time a giant flying squirrel landed on the trunk of the tree just above my head. We surprised each other, but it quickly glided to an unoccupied tree over one hundred feet away. On many occasions I heard Spotted Deer utter their alarm barks as tigers and leopards patrolled the roads bisecting the grasslands. One night as darkness was quickly approaching, a herd of deer began barking about half a mile away. I couldn't see anything, and the light was fading rapidly. More Spotted Deer sounded their warning as the intruder continued its journey along the road that passed right under my tree platform. The tiger was headed my way. I could feel it. The alarm calls were emanating only a short distance from my tree, and I strained to see the shape of a large felid, but it was too dark by this time to see anything. The alarm calls ceased; one minute went by and then another. Then the deer began barking again several hundred feet on the other side of my tree. The deer continued barking for several more minutes before they went back to grazing. The next morning I climbed down to see the tiger footprints in the road just below me. The Spotted Deer had heralded its arrival, but without a night-vision scope all I could do was triangulate the tiger's movements by the proximity of the alarm calls. On other occasions I was more fortunate. One New Year's morning, instinct told me to look down from my platform at first light to see the gray shadow of a large tiger sauntering along the road.

Not all my fieldwork was confined to the reserve. According to unconfirmed reports, there was a surviving population of Blackbuck Antelope in the Bardia area. In Hindu mythology Blackbuck pulled Krishna's chariot across the sky. The spiraled horns and striking black foreheads, shoulders, and backs with a white underbelly give the males a dramatic appearance. At one point Blackbuck were the most abundant ungulate in India. A resident ornithologist in Nepal described a train trip he took across the northern plains of India, where he saw herds of Blackbuck as one might have seen groups of Pronghorn on the Northern High Plains of the United States during the 1800s. The comparison offers other parallels: both Blackbuck and Pronghorn were chased by cheetahs, the Pronghorn by the now extinct North American species and the Blackbuck by the extirpated Asiatic Cheetah, which supposedly still lingers in Iran. The black-and-white markings, so appreciated by nature lovers, became a popular target for soldiers and private citizens. In India, land clearing added to the decimation of Blackbuck populations. By the time I arrived in Nepal, there were more Blackbuck on game ranches in Texas than in all of India and Nepal combined.

Gagan and I walked to the location where Blackbuck had been seen. We found seven individuals living in a cultivated area, taking shelter during the day in a vast patch of marijuana. If you turn over soil in the western Terai to plant a garden, the dominant weed that spreads across the clods of earth is *ganja*. Fields allowed to go fallow become thickets of marijuana. The villagers living near the Blackbuck herd were poor farmers from India who, as devout Hindus, wished no harm to the *krishnasagar*, as Blackbuck are known in Hindi and Nepali. They only wanted them to stay out of their croplands.

Gagan organized the farmers to flush the Blackbuck from the marijuana jungle so I could photograph them. Proof of the small herd could help attract the funding to set up a small reserve or even to pay for a translocation to Bardia once the population expanded. I began to plan a quick publication in a scientific journal, but then I thought, how would I describe the Blackbuck's habitat? As a 2.5-meter-high overstory of *Cannabis sativus*, or what it really was, a field of Nepalese reefer? I could see the headlines now: "Young American biologist finds endangered Nepalese antelope living in field of pot"! I eventually published a short article in an obscure local journal, minimizing the risk to my professional reputation, and we managed to have the farmers' plot formally protected and the farmers compensated.

All volunteers at the end of their service ask themselves the perennial question: what difference did I make? Did I leave behind anything lasting? Most Peace Corps volunteers in Nepal can point to something tangible: a suspension bridge they constructed, a water supply system installed, groves of fruit trees planted. I like to think that I helped put Bardia on the map biologically, that my surveys led to extensions of the reserve (Bardia is now three times the size it was in 1975), and that I contributed in some small way to preserving one of the great nature reserves of Asia.

Like most volunteers who are honest about their experience, however, I realize that I probably took away much more than I ever gave back. I had my first taste as a survey biologist. Professionally, the experience cemented my desire to go on to graduate school to become a wildlife biologist and to eventually return to Nepal. When I crossed the Babai for the last time, heading back to Kathmandu and America, I knew in my heart that my departure was only temporary. A former volunteer had warned me that I might go back to the United States but I would never leave Nepal. Twenty-six years and countless visits later have proved him to be a sage. My improbable friendship with Gagan was very hard to leave behind. I have never again experienced such a deep friendship with someone from another culture.

During one of my last nights in a Bardia treehouse, I looked down one morning to watch large herds of Spotted Deer grazing below me. A tiger came walking by and immediately spooked them, sending up a chorus of alarm barks. I never rooted for the tigers to kill a deer in front of me, as on my first night in Chitwan. I was happy enough to be in a balcony seat watching this drama play out below as it has a million times before: the stalking tiger, the alert deer, the race for cover. For the first time in my life, I felt as if I was a jungle guardian. Even though this wasn't my country or my private reserve, I had spent more time out with these wild animals than anyone before me. It was then that I had an epiphany. My connection with nature seemed so strong that if I were to have died the next day at the age of twenty-four, I would not have felt cheated of a meaningful life. And if that were my fate, then those who loved me, out of respect for all I worked for, could scatter my ashes here, and thus something from my body—a calcium fragment, a bit of nitrogen—would find its way into a fig tree, a langur monkey, or perhaps even a tiger.

The most powerful lesson, though, and one that has buoyed me as an adult in the course of sampling life's menu of personal triumph and tragedy, was the result of my very first experience in Bardia. Or rather, the one on its eastern boundary when my elephant, her two drivers, and I were swept away trying to cross the Babai River. Somehow we got across that river, and I lasted the two years and contributed something meaningful to conservation, even if I never actually followed Dr. Mishra's instruction to census the tiger population. The important thing was that I persevered. I can't recall how many times since I have remembered my promise to myself—that if I managed to get across the Babai on that elephant I could handle any challenge, emotional or physical, that life had in store for me. I often wonder if there are people who live charmed lives in which nothing wretched ever happens and they and all those they touch somehow live in a state of uninterrupted happiness. There may be a quorum of such blessed souls, but for the rest of us who are less fortunate, I learned a priceless lesson that all young people must discover in their own way: when life knocks you off your horse, or in my case your elephant, get back on and cross the damn river.

❧ ❧ ❧

What started in the Peace Corps extends to the present with the inauguration in spring 2001 of the most ambitious wildlife habitat recovery program in Asia: a landscape-scale conservation initiative known as the Terai Arc. The goal of this project is to string together the jewels of national parks and reserves created twenty-five years earlier into an emerald necklace stretching almost one thousand kilometers across the lowlands of northern India and southern Nepal. Using wildlife corridors and locally managed forests as links, we seek to reassemble this magnificent jungle in order to conserve tigers, rhinos, and elephants for posterity. One might describe the list of ingredients needed to reconstruct the Terai Arc as one part Kipling, one part conservation biology, and one part rural development. It is a vision in which we bring back the Era of Great Mammals within the context of the twenty-first century, a grand effort in which villagers and tigers map a new geography that allows man and beast to live in harmony in one of the poorest yet most wildlife-rich regions on Earth.

22

A Mother's Worst Nightmare

TONYA M. HAFF

Have you ever noticed that once you learn a new word, a phrase, a concept, all of the sudden you find it everywhere you turn? That was how I felt about field biologists after becoming one myself: suddenly we were everywhere and anywhere, popping up in tattered clothes at the local bar, passing and waving to each other in battered pickups along dusty roads. And if I wasn't running into someone who also spent twelve hours a day documenting animal behavior through gritty binocular lenses, I was hearing stories of the relatives of friends or acquaintances who were also in the line of counting ticks or ducking angry nesting raptors. A mother's nightmare, really.

One afternoon my heightened sensitivity to the existence of field biologists led me into a long conversation with my dental hygienist, Jan. Jan loved to chat, and somehow between tapping tartar and suctioning saliva, she extracted from me my profession. Her eyes, protected from the spray by a large pair of safety goggles, lit up, and even though a sanitary paper mask covered her mouth, I could tell she was smiling.

Her daughter Diana was a field biologist as well, and she had even worked on several of the same gigs that I had. But by the time Jan was done telling me Diana's latest story, my own stories as a field biologist felt relatively tame. I also felt a lot less sorry for my own worrying mother.

"Do you study birds?" Jan asked, and around the probing steel instruments I nodded affirmatively. Thrilled that we shared a common interest, she began to tell me the story of Diana and the bear.

Diana, a petite and beautiful redhead, got a field job after college studying nesting Harlequin Ducks in Alaska, working with a local game agency. Stationed along Prince William Sound, she was to travel the slow inland rivers and sloughs by kayak in search of nesting ducks. This work was to be done alone in grizzly country, and so Diana was required to be trained in gun safety and to carry a loaded shotgun with her at all times.

Like me, Diana had been educated in the liberal tradition of nonviolence, animal rights, and vegetarianism at U.C. Santa Cruz, and she harbored a strong aversion to guns. Further, she felt that to shoot and kill a grizzly, the essence of wilderness itself, would be a cardinal sin against nature.

"Who am I to come into grizzly territory and kill a bear, even in self-defense?" she wondered. "There are so many more people than there are bears; I'm the outsider here. Better that if I am attacked I surrender my life to Nature. Besides, I'm sure that a grizzly could sense that my intent is pure and would never attack *me*." It was with this attitude that she approached the gun safety class and prepared for her first excursion into the field.

The night before she left on her first trip, Diana called her mother. Skipping the niceties, she broke the news to Jan. "Mom, I'm calling to tell you that tomorrow I'm going to work in grizzly country, and I've decided that if I am attacked, I'm not going to shoot. There's no way I could kill a wild bear and live with myself. So I'm just calling to say goodbye." She hung up before Jan could protest, or worse, start to cry.

In the waning dark of the next morning, Diana loaded up her gear in the white state-owned truck and piled in with the other field crew on the way to their sites. Her drop-off site, not much more than a bend in the dirt road next to a sinuous stream, was several hours into the wilderness. Once there, she loaded food, clothes, radio tags and tracking equipment, GPS unit, first-aid kit, and flagging tape into the narrow kayak. With great reluctance she placed the shotgun on top of the rest of her gear; she would not be allowed to go into the field without it, and since her field supervisor, Keith, was there helping to load the boat, she could see no way to "accidentally" leave the gun behind.

Diana slid the boat, riding low with gear, into the water and paddled upstream. No one had found nesting Harlequins on this stretch of stream for the past several years, and Keith had told her she likely would not find evidence of nesting either. Flipping the oars through the rippled water, she found the solitude both intimidating and challenging. After an hour or so of quiet paddling she arrived at her field site and hauled up into the rushes along the muddy riverbank.

Loading up her field gear in a large external-frame pack, Diana fought through the underbrush of the stream bight. The shotgun added quite a bit of extra weight, and she cursed it aloud. By the end of the day she was tired and disappointed. She had spent close to seven hours fighting vegetation while combing the riverbanks for signs of Harlequin nests or nesting activity, and although she had seen two males, she had found no signs that the birds were breeding along this small, slow-moving stream. Exhausted, she set back toward her boat with the long Alaskan afternoon sun still shining overhead.

Absently thinking about the iridescent sheen of harlequin feathers and the damp smell of mud, Diana approached the narrow bight of the slough where her kayak rested. She was perhaps two hundred meters away from the boat when she noticed a dark, hulking shape take form in front of her. A bear. Its long, grizzled guard hairs glowed, lit up by the same late afternoon sun that had but a moment before felt so reassuring. It was a grizzly, foraging on the dense patches of huckleberries that thrived in the damp, acidic soils near the stream.

A branch snapped under her foot, and the animal looked up sharply, its heavy muzzle stained dark purple with huckleberry juice. Perhaps its eyesight is poor, she thought, as it strained its head about, sniffing. It looked a bit confused and absolutely gentle. Diana took a step back and lowered her backpack to the ground, feeling blindly for her camera as she kept her eyes on the huge but innocent-looking animal in front of her. This was a photo opportunity she just couldn't pass up. As she reached down, the grizzly turned, faced its shoulders squarely toward her, and sniffed the air once more. The metal camera was cold to the touch, but her fingers were already numb from a sudden chill that seized her body. Without warning, the bear lurched forward and charged toward her, barreling the short distance through crashing underbrush.

Diana found herself staring down the barrel of her loathed shotgun, which was aimed between the bear's eyes. She felt her finger grip the trigger and tighten, and as she watched the grizzly swerve sharply to her right, she heard a loud blast and felt a painful kick in her shoulder. She smelled acrid smoke. Someone else seemed to be controlling her body, and as she lowered the gun to her side, she wondered who had shot at the bear. All was silent for a moment but for the ringing in her ears, and then, finally, the *sweet-sweet* of a Yellow Warbler broke the spell, and the air seemed to move again.

Forcing her trembling legs to move, Diana thought about making her way back toward the kayak and her transistor radio, her only contact with the outside world. Her wild shot had missed the bear, which was now somewhere between her and the boat. She would have to walk around to avoid the bear, but that would mean walking in the stream. She waded into the cold, slowly moving water. The willow thickets along the side of the riverbank made navigation difficult, and the deep mud sucked at her boots, slowing the trip to a crawl. It was in this manner, with her still-warm shotgun raised above her head and thus kept dry, that Diana found her first Harlequin nest. As she rounded a bend in the slough, the dappled female fled, scattering dander and down into the air as she squeaked loudly and slid off the nest into the stream. Six eggs, dull buff, shone from under a root burl on the streamside like a beacon.

When she finally reached her kayak several hours later, Diana collapsed into the hull, shivering and wet. Nestled in a waterproof bag at the front of the boat was her transistor radio. Picking it up, she tuned in to channel 51, her crew's channel. After thirty seconds or so of calling in, Keith's gruff voice, tinny over the distance, demanded her whereabouts and her reason for not reporting sooner.

"Sorry about that, but I found a Harlie nest!" replied Diana. She hesitated. "I also saw a grizzly."

Keith's voice, surprised at the nesting news and yet concerned about the bear, asked if she would be willing to return soon to check the nest.

"Sure . . . " she said, as the image of rippling muscle and fur charging toward her flashed through her mind. "I've got my gun." She paused, and then added, "If my mother calls, would you let her know that?"

23

Potheads

SHARYN HEDRICK

The nor'easter had been blowing for two days. Everyone living on Cape Cod was accustomed to enduring the gray skies, driving rain, and steady thirty-mile-per-hour winds that came with these storms. This one was different, though. It was early in the season—barely mid-November—and there was a full moon, not that anyone could see the night-time sky. The combination of the full moon and gale-force winds provided for exceptionally high tides.

My clock radio went off as usual at five-forty-five in the morning, but I was warm and cozy, so I snuggled further into my quilts. I attended Bridgewater State College on the mainland, an hour and ten minute drive across the Sagamore Bridge from South Yarmouth. My first class was at eight o'clock and I didn't have that much time. I half listened to the disc jockey on the Hyannis radio station (the only radio station I could get until I crossed the bridge) as he rambled on about nonsensical news items and played the "Piña Colada" song yet again. Suddenly, I heard something that made me bolt straight up and reach for the volume. He reported a whale stranding in Wellfleet, about ten miles east of South Yarmouth, or "down Cape," as the locals say. Sixty-five Pilot Whales, *Globicephala melaena,* had beached themselves during the overnight high tide and were stranded far up into a marsh area. This was the first whale beaching on Cape Cod since the 1950s, when one hundred whales met their deaths in Wellfleet; during the 1930s as many as three thousand Pilot Whales beached in the same area. The New England Aquarium's Marine Mammal Rescue Team was on site in Wellfleet assessing the situation. Apparently, this pod had tried to strand itself at Cosby Landing in Brewster the day before, but the three that made it to shore were turned around by workers from Sealand Brewster, and they rejoined the pod and swam away.

My feet hit the floor, and I raced to the phone to call Dr. J, my marine biology professor. He was affiliated with the aquarium and frequently went with them on rescues. He told me he was aware of the stranding and said that I should come to school, where we would all meet and decide how to help.

I made record time getting to campus that morning, and I spread the word to other marine biology students as I rushed to my professor's office. He had been in contact with the aquarium people, and it had been decided that the remaining live whales would be euthanized. They were so far up into the marshland that trying to get all of them back into the sea would be impossible. It was too shallow to bring in a barge, and it would require using backhoes to lift each whale, which would not only destroy the fragile marsh but kill the whale in the process. The massive weight of the whales would crush their lungs regardless of whether they were moved or not. The rescue team tried to save a few of the smaller whales near the surf line, but as soon as the whales were back in the water they turned and rebeached themselves. Rarely can a beached whale be saved. Along with their sheer weight pressing on them, their blood circulation is affected and exposure to the sun raises their body temperature. The whales were suffering, and euthanasia was the humane answer. We were devastated, of course. However, we would be allowed to go to the area and retrieve one or two of the bodies for study if we could get them off the marsh.

The next morning, five students met Dr. J at school and drove to the Cape. Since I already lived there, I drove directly to the site in Wellfleet. The whales couldn't have picked a worse spot to beach. A few scientists theorized that the extreme high tide and the storm had caused a miscalculation by the pod's leader. Jeffrey's Point, a spit of land that is always submerged, but even more so because of the high high tide, separates Wellfleet Harbor from Cape Cod Bay. Instead of staying on the bay side of the point, they veered to the right and ended up in Wellfleet Harbor all the way up to Lieutenant Island. The marsh, Blackfish Creek, which flows into the harbor, and even the narrow one-lane wooden bridge connecting Lieutenant Island with the shore were all under water. The whales didn't stand a chance. The marsh was quite large and belonged to the local chapter of the Audubon Society. The society was naturally concerned about any damage to the marsh. When we arrived we learned that local residents were also worried about having the huge whales wash up on their private beaches. The manner in which the bodies would be disposed of was unpleasant, but because of the remote area it was the most practical. The Audubon Society agreed that all the whales should be cut up and dumped into the marsh tidal pools to decay naturally.

Pilot Whales are considered medium-sized—about six meters in length—and they weigh about three tons. They feed exclusively on squid, resorting to herring, mackerel, and capelin only when the squid has been exhausted. They are solid black with a faint anchor-shaped patch on the belly, which can be seen only from underwater. Their heads are very bulbous, owing to the large rounded "melon" on top. This gives them their common name: potheads. I had seen these potheads from a boat, but I'd never been as close to them as I soon would be.

The nor'easter finally blew itself to the Maritimes and left the Cape with spotty sunshine and diminishing winds. Dr. J and the crew arrived, and the resident Audubon wildlife manager escorted us in trucks to the stranding site. He parked on the side of the narrow road a fair distance from the whales, and we walked the rest of the way, winding single file around grassy hummocks and tide pools.

I cannot fully express the emotions that washed over me as I approached the dead whales. It was horrifying. At first sight, they looked like enormous black Goodyear blimps dotting the marsh, some lying half in tidal pools. They didn't look real. The impact of seeing all these dead creatures didn't fully hit me until much later, when I was back home.

We all slowed as we reached the whales, standing in silence as we surveyed the disaster. We were told, quietly, to pick a whale and start cutting. There was concern that we wouldn't finish before the next high tide. It was eight-thirty in the morning, and we had until six-thirty that evening. In order to obtain our two whales for school, we had committed to help dissect and dispose of all the bodies.

Someone handed me a flensing knife and a butcher knife. I stood looking at them as though I'd never seen a knife before. I was supposed to use these? These tiny, seemingly delicate blades? And what was worse, I was supposed to cut into the body of one of these poor creatures. Not quite the same thing as cutting open a dogfish in chordate anatomy lab. A friend called out and asked if I was okay. I nodded and walked over to a small whale nearby. It was a juvenile, only ten feet long. I was wearing black fisherman boots that came to my knees, old jeans, and a kelly green foul-weather jacket. I knelt in the soon-to-be-bloody grass beside the young whale and ran my bare hand over its skin. It didn't feel like rubber as I had expected, and it didn't have the sand-papery finish that many sharks have. It was cold and smooth and strangely unlike anything I'd ever felt. Years later I had the opportunity to run my hand over the skin of a humpback whale as it came up along-side the boat I was in. It didn't feel the same.

I sat there on the cold marsh grass in the biting wind, the flensing knife in my right hand, not having a clue where to begin. Dr. J apparently noticed my perplexed state and came over to show me where to make the incision. "Start in the midline near the tail and in one sweep cut upward to the sternum. Oh, and use the butcher knife. The flensing knife is to separate the blubber from the body once it's open." He returned to his own whale.

Okay. That sounded simple enough. I held my breath, closed my eyes and plunged the knife into the whale's body. It broke—snapped off at the handle. *Oh, great,* I thought, *Dr. J is going to think I'm really an idiot.* Luckily, Ted, our teaching assistant, saw my predicament and brought me another knife. He was also kind enough to start the incision for me. With the new knife in my hands I gradually began dissecting the whale. It seemed to take forever. Hours went by—it had to be hours. My arms were beginning to ache, my fingers burned, and I was working up a sweat even in the cold air. Finally, thank God, I reached the sternum. I sat back on the grass and shook my arms to get the circulation moving properly again. Then I leaned into my work with gusto. I was astounded at the thickness of the blubber, four or five inches, and this was only a calf. Using the flensing knife, I was able to flense away the blubber with more ease than the original cut.

I stopped for a breath and looked around to where the others were working. Everyone was quietly cutting up the whales. Occasionally, someone would ask a question or make a comment or call out for a sharpening stone, but mostly we worked in silence. We'd been here two hours already, and there was so much more to do.

I turned back to the youngster lying before me, split from stem to stern. I took another deep breath and began again. After stopping numerous times to sharpen my knife, I eventually got the blubber removed and sank it in the shallow tidal pools nearby. Small crabs and other benthic fauna quickly discovered the feast in the ponds and scurried to get their share. Gulls circled noisily overhead, ready to swoop in and steal a morsel. Nature was recycling.

I was ready to remove the organs now. This is when I fully realized what I was cutting up. Whales, being mammals, are incredibly similar to human beings. They have the same intestines, stomach, liver, kidneys, and heart, and two enormous lungs. The lungs reminded me of an accordion-pleated skirt, the ridges were so evenly spaced. They had a bluish tinge to them as well. I removed them and then the heart.

I was wearing slightly oversized rubber gloves, but there was so much blood that my green jacket sleeves had turned an ugly dark rust color. Blood also seeped into my gloves at the wrist, making a squishy noise as I moved my hands. It was nearly noon, and the sun had managed to beat away the clouds. The temperature had risen to the upper fifties, which was nice for November. I took off the jacket, pushed up the sleeves of my sweatshirt, and stuck my arms back into the cavity of the whale. I remember thinking that it should be warm inside the whale. Instead, all I felt was bone-chilling cold. Lying in the marsh, the whales lost their body temperature as quickly as a human would.

Dr. J shouted to everyone to be sure and find the hyoid bone. He wanted to rearticulate the whale back at school in its entirety. I had no idea what a hyoid bone looked like, but I kept searching around inside. By this time the shock of first seeing the whales had worn off, and we were all a bit more relaxed. We were also getting tired. With tiredness came frivolity, and we grew giddy. As I probed around inside my whale looking for the hyoid bone, my hand gripped an organ that left no doubt as to its purpose. I gulped and called out to Dr. J, "Are whales *exactly* like humans?" He raised his head in my direction and being a man of few words, said "Yes," and then returned to his own whale's innards. "Well," I said, "I've got a male."

Every head on the marsh lifted and looked in my direction. It was dead silent, and then the giggling began, which soon turned into full laughter. I wouldn't live that down for a very long time.

Once the whales were reduced to skeletons, the heads were removed on the two we were taking back to campus with us. This was done by disarticulating the head behind the blowhole. It was then roped to a pole and taken back to the truck slung on the shoulders of two men, the head swinging back and forth, smiling grimly, as they tromped through the marsh.

A few reporters showed up, slipping past the Audubon defenses at the Lieutenant Island bridge, and began recording comments and taking pictures. The wildlife manager who had escorted us in waved them away saying, "No. We don't want anybody here. No pictures. This isn't scientific; it's just butchery. It's no good to have people see this."

Unfortunately, one reporter didn't leave soon enough. He heard a comment I made to some of the others carrying pieces of whales and used it in his article. I said something to the effect that the pieces I was carrying were "spare parts." I wasn't being flippant, but that is the way he reported it. There were three colleges—University of Guelph in Toronto, College of the Atlantic in Maine, and Bridgewater State College—involved in the removal of the carcasses, and each school was allowed to take a whole whale or pieces of one. We all carried out sections for study, and we were all very, very tired. There was so much blood everywhere. The mud in and around the tidal pools was pinkish-gray, and the surrounding yellow marsh grass was matted and stained a deep red. All of us looked like we were part of a MASH unit, blood-spattered and dirty. Humor was the only way to keep our sanity after spending hours knee-deep in cold mud and guts.

We finished by four-thirty with the tide at our heels. By the time I reached home, my body's defenses were exhausted; I was frozen and shaking, even with the car heater turned up high. I think the shock of the day's events had finally taken its toll. With blood on my clothes, my arms and face, and in my hair, I looked like I'd been in a car accident.

I stood in a hot shower for a very long time.

24

Gumbo

LYNN SAINSBURY

I cut my teeth doing fieldwork on the Black Hills National Forest, South Dakota, in 1987. The backbone of the Hills is a massive block of coarse-grained granite—rock that shoulders the famous presidential faces of Mount Rushmore. The granite weathers into chunky sands that drain well and don't create conditions for mud. There are places, however, where thick sheets of clay overlie the country. With even a little rain the clay transforms itself into sticky goo commonly called *gumbo*. It doesn't matter what kind of rig you're driving or what kind of tires you've got on it, gumbo will take you down. Monster trucks wallow helplessly in the stuff while their drivers cry like babies. Gumbo always wins.

Though we both had fresh biology degrees, my work partner and I were hired as range techs. One of our tasks was to count cows coming onto the Forest. The district was broken into several allotments. On each one a certain number of cow/calf pairs were allowed to graze for a certain number of months. The ranchers paid for these grazing rights based on the allotted number of cattle rather than the actual number. There was no incentive to have fewer cattle on the Forest but great incentive to slip a few extra on. It was our job to make sure no more than the allotted number poured from the trailers into the pastures.

Here we were, a couple of fresh-faced, early-twenty-somethings standing in the back of a Forest Service puke-green Dodge pickup in the middle of a tide of brown-and-white Hereford cattle. We split the tide; I counted cows passing on one side, my partner counted the other side. Neither one of us knew doodley-squat about cattle, and the ranchers all knew it. They tolerated us because they had to. Their interactions with us amounted to handing over a piece of paper with a number scrawled on it which we would sign. Later they would bring it in for our boss, Mike, the range conservationist, to slip into a file somewhere. In reality it was impossible to make an accurate count of the multifronted crowd of bawling bovines that buffeted our truck; we didn't really know how many cattle went on the Forest, but we always signed the paper.

It was during one of these cattle-counting trips that we first encountered gumbo. In the early afternoon we met a rancher who was transporting his cattle in a variety of beat-up old horse trailers pulled by a fleet of worse-looking flatbed pickups. As the unloading commenced, so did the rain. It was a typical summer thunderstorm with drenching showers and plenty of lightning. We stood slump-shouldered in the back of our truck, "1551," and desperately tried to count the thunder-spooked cattle that roared past.

After the cattle were unloaded and the paper signed, we shrugged off our soaked-to-the-skin stupor and began preparing to leave. The very last thing we wanted to do was prove that we were a couple of know-nothing greenhorns to ranchers who already knew that's exactly what we were. Problem was, we had parked the truck at the bottom of a steep two-track road, and though she sported good tires, 1551 was still a two-wheel drive. While the ranchers stood around bullshitting, we piled into the rig and started up the hill. We made it about fifteen feet before spinning out. We backed down to the flat again and gave her another go with exactly the same result. We foolishly tried this a couple more times before coming to the obvious conclusion that this feat required chains. At the bottom of the hill we jumped out of the truck and started rummaging around the bed for the tire chains, trying not to make eye contact with the ranchers, who were all watching us—eyes twinkling in mirth.

The spinning tires had managed to throw about a hundred and fifty pounds of clay up into each wheel well, effectively closing the gap between rubber and sheet metal to about three-quarters of an inch. Gumbo also wrapped around the tire, making it twice as wide as normal. Neither shovel nor Pulaski could coax all that gumbo from the wheel wells and tires. We grubbed around with our fingers, finally getting enough off that the chains could go on. By this time every part of our bodies were covered with wet, sticky, gray goop. It was mashed into our hair and plugged our ears; it was four inches thick on our boots and had transformed our hands into slimy paws; it packed the space between shirt sleeve and arm so that we could barely bend our elbows. But we got those damn chains on, waved sheepishly to the ranchers who were still just standing around watching us, tried to scrape as much gumbo off ourselves as possible, clawed our way into the truck, and tried it again. With the chains on we made it about twenty-five feet. Though the chains hadn't improved our traction, they did provide a much better surface for the gumbo to latch onto, and the wheel wells became totally packed once more.

We climbed out of the truck again, and our fallen faces must have sparked a pity response in a couple of those hardened cowhands. While we walked around the truck considering options, one of them came up, shook his head, and said, "Pretty awful stuff, ain't it?"

With much head bobbing, we agreed.

"Why don't you just follow us down to the road?" he asked.

Down to the road? We knew there was a good gravel road about a half-mile farther down the hill, but between the road and us were a pasture and a barbed-wire fence with no gate in sight. The one thing we did know was that it was nearly a hanging crime to cut through a fence. We would have left 1551 sitting in the gumbo and hitched back to town before we'd have cut that fence.

While the ranchers closed up their trailers, I tried beating my feet against the bumper to dislodge the gumbo. One of the guys came up and handed me a stick. "Works better to scrape it off," he informed me.

They all climbed in their pickups and started snaking down the hill through the field. The trucks slid one way and the trailers another, but we all made it to the fence on the other side without too much trouble. The guy in the first truck climbed out with a pair of fencing pliers, cut the wire, and let the rest of us drive on through. As we were pulling onto the road, the man who'd offered me the stick called out, "Run them chains down the road till the gumbo's off 'em; that's the easiest way to get 'em cleaned up. And make sure you get it offa the truck afore it dries or you'll never get it off, and you'll catch hell from Mike!"

We spent the rest of the day at the car wash.

25

La Yunga

BETSY L. HOWELL

As a child, I fantasized about being lost in the wilderness. Daydreaming for hours, I would build my own cabin, fish, hunt, gather wild foods. I watched movies that glamorized the perils one might encounter after a shipwreck or some other calamity. In one a survivor snagged a duck by hiding behind a floating log. In another a man escaped a polar bear's attack by squeezing into the tail section of a crashed airplane. Such daring and bravado! I was sure that I too could survive in lands where weaker souls would curl up and die. Growing up in suburban Tacoma, Washington, I wondered if I'd ever have such an opportunity. In early 1993, while serving in the Peace Corps in Argentina, I got my chance.

My assignment was a research project on pumas in the western province of La Rioja. A national research organization wanted to assess puma depredation on small family ranches. People eking out a living from their goat herds claimed pumas killed many of their animals. The herders, in turn, killed pumas. My job was to develop a proposal to determine the extent of predation, devise strategies whereby goats and pumas could coexist, and learn about the biology of the cats. As I sat through the twelve-hour bus ride from Buenos Aires to my "site," I felt thrilled and scared. Thrilled because my other childhood dream of working with wildcats was coming true. Scared because I had never done anything like this. Aside from reading a lot about the puma and writing my senior paper in college on a population of cats in central Idaho, I had no experience with these animals. After a few weeks in La Rioja I knew I needed practical training.

I learned that a man named Pablo Peruvic was studying jaguars in the northern Argentine province of Jujuy. I called him from the research station, the phone crackling as if a pack of mice was having a feast on the line.

"Señor Peruvic, soy Betsy Howell, y…uh…y quiero hablar contigo sobre tu proyecto de jaguar." The young clerks in the office smiled at my fractured Spanish. I strained to hear while my cheeks turned red.

Pablo sounded younger than I had expected. He spoke quickly, and I caught only words and phrases. He said he was working in "La Yunga," a landscape of semi-tropical rain forest and high mountain meadows.

"We are looking at depredation by *el tigre* and putting radio collars on captured cats," he explained. "In the summer, during the rainy season, two or three jaguar live in La Yunga. I will be going to the study area for ten days soon." Pablo said something else about a trap, or tracks—I couldn't tell. Our connection was fading.

"That's great!" I yelled in English, then remembered where I was. "Me gustaria acompañarte en La Yunga y aprender sobre los gatos grandes." I knew I sounded like a child, but that was my best Spanish. Pablo agreed to my coming, but the line died before I could find out exactly when I should arrive. I had been in La Rioja only a couple of months; should I ask for leave so soon? I explained to my supervisor, Pedro, that I needed training. He couldn't have cared less. My arrival in La Rioja was similar to someone in the Australian outback finding a whalebone; no one knew why I was there. My co-workers believed it was silly to study "the puma problem." You killed pumas when they took goats. What was the problem?

Pedro told me to go. "*Vaya, vaya*, and have a good time!"

After a thirteen-hour bus ride to Jujuy I found Pablo's home atop a hill that overlooked a city stratified into the very wealthy and very poor. This province abuts Bolivia and has for many years provided refuge for people escaping that country's economic crises. Pablo lived with his mother, looked to be about thirty, and welcomed me into their home, where he insisted I stay while he made preparations for the trip. For two days, I tagged after him as he organized the ride to the drop-off point, gathered supplies to repair and monitor the jaguar trap, and purchased enough food to last ten days. Pablo spoke little. He usually had a wad of coca leaves in his mouth so when he did speak I had trouble under-standing him. He was thin with wavy hair cut short against his scalp. Pablo ate only one meal a day and moved like a rushing river, fast and deliberate. He told me about the study area.

"During the summer there aren't many people in La Yunga. It's too rainy, and there is plenty of grass for their livestock close to home. We'll be alone in there. It's a four-hour hike into the *campamento*." He glanced my way quickly as if sizing me up for the challenge. I tried to look strong. Though I had spent numerous days in the field working as a biologist with the U.S. Forest Service in Oregon, I had never spent this much time in the wilderness. But I had backpacked before; how hard could this be?

"Are there trails?" I asked.

"There is a trail to the *campamento*."

The next morning, after borrowing a tent from Pablo's dad and a car from his grandmother, we set off, driven by Pablo's friend, Rodolfo. The pollution and shantytowns of Jujuy became memories as the car took us higher into the mountains. Every shade of green blended across the land, like a series of paint chips. The sky broadcast a deep azure, and the sun warmed us even as the temperature dropped with the increasing altitude. At the drop-off point we said goodbye to Rodolfo, who also had a cheek like a ground squirrel, stuffed full of coca leaves. Rodolfo gave Pablo some coca and then offered me some. I declined.

"It will keep you from getting hungry," Rodolfo winked. In the few days I had been with Pablo, I had not adapted well to his one-meal-a-day plan and felt ravenous most of the time. This had not gone unnoticed, apparently.

Into the wilderness we strode, carrying two backpacks containing one tent, two tarps, two sleeping bags, one hammock, chicken wire and newspapers (for the trap), the medical kit for tranquilizing *el tigre*, and our food. I also packed raingear and enough clothes to last a month. Pablo carried a machete in his right hand and pronounced, "Aquí está La Yunga!"

We had not walked five minutes when we encountered a small log house.

"This is the home of Don Ocampo," Pablo said. "We need to buy a chicken from him for the jaguar trap."

Pablo spoke with Señor Ocampo, and then we waited while his wife chased after the chicken. Finally, we started along the trail that would lead us into the dense forest. The sounds of civilization vanished. Only the caterwauling of the wild forest turkeys high in the treetops and the swinging shadows of monkeys broke the still quiet of the land. I stopped to look at everything with my binoculars, but Pablo walked far ahead of me, oblivious to anything but reaching the camp. I expected a lunch break. What I got was a short rest and Pablo's offer of a few hard candies. I asked him about the trap.

"It's six and half feet long and three and a half feet high," he said, "and made of wood. I carried it in pieces, which took three trips."

"Where is it?"

"Close to the *campamento*, about twelve kilometers from here."

We continued, Pablo carrying his machete in one hand and the chicken in the other. He leaped across the slippery, mossy rocks of a small river in his tennis shoes like a forest fairy. I gingerly stepped from rock to rock, hoping I would not go tumbling in, along with my camera, books, and binoculars. After a while, I offered to carry the chicken, which had, up until this time, sat passively in the crook of Pablo's arm. She must not have cared for my uncertain fumbling as I held her. Near a large, dark pool the beast flapped out of my arms. I cringed. We were a long way from Don Ocampo's house, and there would be no jaguar without live bait. The hen flew to the other riverbank, clinging to a rock wall. Pablo waded across. The bird flew higher. He started climbing, and I came from the opposite direction. Slowly, we closed in on her until Pablo pounced on the squawking señorita.

The *campamento* was a semi-open area near the river with two accommodating trees for the hammock and forest furniture Pablo had constructed on previous visits. There was a small table and a two-shelf cupboard for silverware and pots, both made of similarly sized sticks tied together with rope. A tripod of tree branches stood over the fire ring, and a five-foot log split in half served as the pantry for vegetables, fruit, bread, coffee, and tea. We set up the fire-red tent and strung the blue tarps over the kitchen area. The *campamento* didn't blend into the forest very well, but it was both rustic and comfortable enough to meet my expectations for wilderness living. Later I swam in the sedate pools of the river. The water temperature brought goose pimples to my skin but felt like an energizing tonic in the weighty air of the jungle.

The chicken awoke us at five the next morning. I was not pleased and suffered no guilt about its impending fate in the trap. Spears of sunlight stabbed their way to our campfire as we drank coffee and ate a spartan breakfast of bread and honey. We soon left for the trap.

The cage was a long rectangle with one end divided from the rest by chicken wire. Into this area Pablo shredded newspaper and installed the hen at her post along with some feed and water. At the other end was a door, held above the entrance by a thick string, which was attached to a false floor next to the chicken wire. When the cat stepped on the floor, the string would release the door and the animal would be caught. Pablo attached the chicken wire only nominally so at least the jaguar would get a free meal.

He then explained the daily routine to me.

"We'll check the trap each morning. If there's a jaguar, we'll sedate it, take measurements, and tag it. If not, we'll spend the day hiking the study area, looking for tracks and kills of *el tigre* and its prey."

"What about lunch?" Everything sounded great to me. My only concern was that I hadn't seen Pablo pack any food for the day.

"No importa," he answered casually, walking off into the woods.

"¿No importa?" I mumbled. "¿No importa?"

Lunch was always important in my book. I stood chagrined for a moment, knowing that I often became lightheaded if I didn't eat regularly. I pondered the food back at camp. Well, I thought, if I faint, Pablo will have to deal with it. I concluded that this quiet, determined man not only revered the jaguar but wanted to be one as well, eating once a day, moving lithely, speaking little. I decided if he could be a jaguar, I could be a puma, also an inhabitant of La Yunga. For our ensuing field days, I packed only a few candies and one piece of fruit, and surprised myself by feeling energetic enough to hike at least ten miles each day.

Occasionally, we used animal trails, but more often we bushwhacked, Pablo and his machete in the lead. I learned quickly that when Pablo indicated a route that appeared impossible to me, it was normal and doable to him. That first day, we hiked up the river and came to a thirty-foot waterfall. Ferns and dew-heavy mosses hung languidly from the perpetually wet rock walls surrounding the falls.

"Oh-oh," I muttered, "¿dónde vamos?"

Pablo looked at me, surprised. "Aquí, no más," he said, as if I had just asked a very stupid question. Here, no more? I thought. What does that mean?

I gazed at the vertical walls, which said to me, "Go through the shrubbery," but which, I think, said nothing to discourage Pablo. We went our separate ways. He beat me to the top but had the grace not to gloat.

As the days went by and more of La Yunga passed beneath our feet, my faith in Pablo grew. I followed him like a puppy. Marveling at the raucous turkeys, the dainty pointed tracks of the Corzuela Deer, and the rounded paw prints of the Ocelot, which supported tiny ponds of rain-water in each depression of toe and palm, I left the navigation to Pablo. In my journal I wrote of our wanderings and sketched the *campamento* and a black-and-white mushroom we found that was as tall as our knees and looked like a lady's garden bonnet. Pablo smoked and chewed coca and seemed to watch everything. I decided I could trust anyone who made forest furniture and survived a day of jungle hiking on three candies.

After our first day of fieldwork a summer storm moved in, opened up a sky bursting with water, and did not leave for nine days. Nearly everything we brought got wet. The hammock mildewed as it hung unused. Inside the tent the edges looked like a beach after the outgoing tide, and my hands and feet shriveled like dried prunes. Only my journal stayed dry, tucked inside a Ziploc bag. Each evening we returned and put our sweaty field clothes by the fire to steam. Each morning we awoke to soaking jungle vegetation and overflowing creeks. The limpid pools I swam in metamorphosed into swirling maelstroms of chocolate-colored water. Only the thought of a jaguar in the trap motivated me out of my sleeping bag. As the days went by, however, and the chicken lived, I became disheartened. I told Pablo the turn in weather was my fault.

"I'd been thinking how perfect everything was. Now look at the situation," I said, during one day of torrential rain when we decided to stay in camp. We were playing tic-tac-toe in the dirt.

"You have to experience all the aspects of La Yunga," Pablo said, philosophically.

Our last day began in the usual manner. We ate bread and honey for breakfast and drank coffee. The jungle, a tapestry of fog strands, was eerily quiet except for the dripping trees and excited river. For several days we had not heard any birds, and we had yet to find any sign of the jaguar. Pablo was even more quiet than usual. I packed into my pocket five candies, one apple, and a flat, hard candy bar that looked like a Big Hunk.

"We'll leave the river today and go up to the high meadows," said Pablo.

I nodded gravely. My excitement for this jungle adventure had mutated into a dull ache of sore muscles, nettle rash, and bug bites over my entire body. My feet never dried out, and I worried about fungus. Tomorrow, however, we would be back at Pablo's house, dry and warm. The thought comforted me.

Drizzle fell all day. Toward early afternoon we left the canyons below and emerged into the pale sunlight of the meadows. We followed the narrow animal trails through the swaying grasses and examined the ground for tracks. Though the low clouds still drizzled on us, the presence of the sun helped immensely. Pablo smiled and seemed more cheerful in his reticent way. In one meadow a male Corzuela raised its head abruptly as we interrupted its lunch. Everywhere life seemed in motion again after the closeted feeling of the constant rain.

Pablo and I waded through the meadow grass all day. Monk parakeets crackled their conversations from the treetops. I saw a black and yellow songbird, which proved to be the *rey del bosque*, "forest king," or Black-backed Grosbeak. I was watching for more birds when I crashed into Pablo's back. We both teetered, trying not to fall forward.

"¡Mira! ¡Mira!" he shouted, jumping off the trail so I could see. There in the mud was one perfectly outlined jaguar track. It was bigger and rounder than the puma tracks I had found in La Rioja.

"¡El tigre!" Pablo yelled, grabbing me for a quick waltz around the meadow. "It's only just passed," he continued excitedly. "Look, there is no water in the toes or palm!"

I snapped photos while Pablo took measurements of the track—a front paw, four inches long and just as wide. I felt ecstatic. This foray into the wilderness, though uncomfortable, had proved worthwhile. Pablo had one more piece of data to fit into the puzzle he was constructing of La Yunga's summer jaguars. We descended the slippery slopes toward camp. I looked forward to the hot meal of spaghetti we had planned, and Pablo seemed content simply to have found a trace of the elusive jaguar.

We had at least a two-hour hike back to camp. As we traveled downward, the heavens closed in above us, and I was amazed that we had climbed so high. At last the river grew louder. The section of stream we came to did not look familiar to me, but I had felt adrift all week in the maze of La Yunga. Pablo cantered downstream as if he knew exactly where we were, so I wasn't worried.

After an hour we arrived at a fork in the river, and Pablo turned in a circle, looking puzzled. It was almost dark, and rivulets of rain ran down my face and neck. Cold and tired, I didn't like the look on his face.

"Let's cross above the split," he said. The river was high and fast, but we had to do it somewhere. The water surrounded me and breached my rainpants. In a few seconds I was even wetter than I had been before, if that was possible.

We continued downstream. We walked and walked and dusk came and went, and by now it was clear we were lost. Neither of us had a flashlight or compass. My irritation rose in the wake of ten days of tiredness, nine days of rain, and now nearly one full day without a meal (I did not consider our breakfasts of honey and bread complete meals). I trudged behind Pablo, hoping to see the red tent at any minute.

Eventually, we came to an old rutted road. Since the study area had no roads, this proved we had gone astray. I consoled myself with two thoughts: one, if we were not at the pickup site tomorrow, Pablo's *compañeros* would look for us, and two, despite its overgrown appearance, the road might lead to someone's house. Pablo looked vexed, his lips narrowing into tight lines. I think he was embarrassed, but I hoped he didn't feel too bad. My lifelong dream of being lost in the wilderness was coming true. I just wished I could enjoy it on a full stomach.

We crossed more creeks. The rain streamed constantly down my face, droplets cascading from my nose and chin. At last Pablo spoke.

"We need to find shelter." He said this as we came to the end of the road. It terminated in a river, wider than any I'd seen so far. It looked to be forty feet across and we couldn't tell how deep. We had passed nothing that resembled shelter, so we had to cross. I held my backpack above me as the water rose to my stomach. The river tugged and trembled. I was just another bulk, like a fallen tree, to be moved along if possible, or skirted around if not.

We made it to the other side, and Pablo pointed to a boulder the size of a large garden shed. "Maybe in there," he said, hopefully.

I didn't know what "in there" meant, but underneath the behemoth was a crawl space, dug out by animals. It was six feet long and three feet wide. I patted the ground inside. The back felt dry but the front suffered rainwater, sliding down the rock's smooth face. Sighing, I agreed.

"We'll build a fire," said Pablo. He seemed cheerful now, as if he'd planned this all along.

"A fire!" I shouted. "With what?" He smiled as if I was a very simple creature. Pablo kneeled down.

"Get on my shoulders, then snap the dead ends off the branches of this tree. They'll only be wet on the outside." This seemed a good idea. We walked around like two circus performers, me breaking twigs, Pablo holding onto my legs. After collecting three handfuls, we worked on the fire inside the burrow. For once, I was glad Pablo smoked, as he had many lighters. But in the end it didn't matter. It was like trying to start a flame inside a sauna, a cold sauna. For an hour we labored. Pablo finally left in disgust. I stared stupidly at the Lilliputian sleeping quarters. Where would we have slept if we had gotten a fire going?

Soon I heard the ringing of the machete. Pablo returned to the cave with a load of palm fronds. We put some inside for bedding, and piled the rest along the entrance. Pablo gallantly offered me the dry portion. Maybe the night wouldn't be so bad.

Using my backpack as a pillow, I assumed the fetal position. An image appeared in my mind of the boulder shifting and the two of us vanishing into the earth forever. Pablo settled into the front and piled up the fronds to block the rain and wind. The air soon stank with sweat, breath, drenched clothes, and the damp underside of the rock. I thought of those Hollywood movies, the survivors in their weatherproof cabins with luxurious bear and reindeer skins to keep them warm. "Humph," I muttered to myself.

A few hours passed. My legs went to sleep and my hips ached. Pablo had the nerve to snore. How could he sleep? I felt like an intestine in a wet body cavity. My claustrophobia kicked in.

"Hey, Pablo, I can't sleep. Let's switch."

He wasn't happy with this idea but obliged me anyway. The front wasn't much better, but at least I could breathe and stretch out. Pablo squirmed behind me.

"¡Che!" he said, using the Argentine equivalent of "Hey!" "No hay nada para hoja aca." ("There is nothing for leafage here.")

Pablo's surprise that the bedding in the back of our shelter was not as good as the front caused me to become hysterical with laughter. According to his assessment, the only thing wrong here was that we didn't have enough leaves! Here we were, sandwiched under a boulder like two earthworms, without an adequate bed! Soon I had tears running down my face. Pablo said nothing, but I could feel him staring at me. I knew I should explain to him that I hadn't gone crazy, but I didn't have the energy to do so in Spanish. He took some of my leaves, lay down, and was soon snoring again. I stared at two lightning bugs that were moving up and down grass blades only a meter from my face. Every time I woke up during the long night I saw the bugs' glowing pinpricks of green light.

The rain stopped by morning. We emerged from the burrow stiff and tired. I looked around sadly at the prospect of no morning coffee. My stomach growled. Rummaging in my backpack, I found the twisted Big Hunk, and Pablo and I split it. I saw that the road that had ended so abruptly on the other bank of the river continued into the forest on this side.

"We'll just walk along this road," he said. "There's no use going back, and I'm sure this will lead somewhere."

His vague conclusions did not give me comfort. Of course, the road would lead somewhere! But I did not have a better idea, so I followed. For the next eight hours we walked in silence, not recognizing anything. We saw no people, nor any signs of humanity. Pablo did not seem affected by the lack of food or sleep. I walked slowly, trying to conserve calories.

Late in the afternoon we came to a four-way intersection. Pablo smiled and said we had passed this way driving to the drop-off point.

"Viene Marco por este lado," he said happily.

I eyed Pablo with suspicion. His friend, Marco, was supposed to pick us up today at Don Ocampo's house. Was Pablo saying Marco would be coming this way only so that I wouldn't kill him for getting us lost? I had trusted Pablo, but he wasn't infallible. I searched his face for uncertainty but found none. Pablo sat in the middle of the road, Zen fashion, and traveled somewhere else in his mind. I opted for the soft dirt of the ditch, falling asleep instantly in the warm sun. After forty-five minutes Marco drove by, the only vehicle we saw all day. He was surprised to find us twenty-five kilometers from Don Ocampo's.

The story would have ended with our safe return to Jujuy but for the matter of our gear and the stranded chicken. The next day Marco, Pablo, and I returned to the *campamento*. The rain had begun again. We now waded through water up to our thighs. After packing up the tarps, tent, and hammock and shutting down the trap, we headed out. The river had come up several inches in only an hour. The water now approached my waist. Pablo stepped on a smooth rock on the riverbed and the current carried him fifteen feet downstream. The chicken screamed. As I crossed a second shallow creek, my left foot slipped and I fell onto a pointed stone, my right knee absorbing the impact of my weight. Every other ill forgotten, I howled in pain, and Marco and Pablo carried me to the other side. I could not walk, and we still had five miles to the truck. They took turns helping me on the long, slow hike out of La Yunga as I silently cursed Hollywood and its fraudulence.

Surviving in the wilderness was not glamorous.

26

Rollie's Catch

CHRIS SMITH

It seems that anytime you put two Alaskans together, the talk invariably gets around to bear stories. Wildlife biologists are no exception. They've got tales about drugged bears that started "coming to" in airplanes or bit right through the tailgate on a Fish and Game pickup. There's one account of an orphan grizzly some pipeline workers sent to town that got into a highly colorful wrestling match with a couple Fish and Game biologists in the paint shed out back of the Fairbanks office. The best one, though, has to be about the time a park ranger by the name of Rollie took on a brownie on the Brooks River with ultra light tackle.

The Brooks is quite the river, with several hundred twenty-eight-inch rainbows and a couple thousand salmon clogging its mere one-and-a-half-mile course. Being in a national park and having a tourist lodge on its banks have given some fame to this stretch of water. Of course, the river's been well known and heavily fished for generations by a healthy batch of some of Alaska's biggest brown bears. Some of the bears aren't too tolerant of those "lower 48" fly fishermen and simply slip off into the bushes when the annual nimrod migration begins. On the other hand, some bears have figured out how to take advantage of the new situation. One clever sow in particular, known to the local Park Service and Fish and Gamers as "Sister," caught on right away that the shouts of a human holding a bent rod, coupled with the buzz of a drag and splashing, meant an easy lunch. Hell, it was a lot simpler to pounce on a salmon while it was concentrating on peeling line off a spool or to just bluff-charge some easterner out of his newly landed catch than it was to spend hours standing at the riffles or falls waiting for a sockeye to make a love-inspired fatal dash through brown legs in white water. Besides, the bears hardly ever got to taste the sweet, firm flesh of a rainbow unless they stole it. (Or waited downstream from some heavy-handed "kill-and-release" type.)

So in '79, Sister got to be quite the nuisance, lying in the bushes listening for the telltale cries of "fish on!" as her cue to give another angler a thrill bigger than the trout he lost yesterday, and giving Superintendent Blinn another gray hair. Knowing that it was only a matter of time until some fool refused to yield his trophy, thus taking the "bluff" out of Sister's charge, the rangers decided it was time to modify this bear's behavior.

I had previously advised them of how we "educated" bears in King Salmon and at McNeil River with a load of 7½'s square on the rump at about forty yards. It usually doesn't take two lessons to get the point across. So the Park Service set out to give a whipping to Sister, and just to make sure she knew what it was about, they decided to stage a fish-stealing incident. Any good lawyer woulda called it entrapment.

On the appointed day Ranger Rollie left his green and tan suit with badges and nametag in his cabin, dressed in his tourist best—even borrowed one of those felt hats with the shaving brush on the side from some German fellow—took his favorite graphite from the rafters, and headed for the Trout Hole (one of the surest spots on the river for any species of fisherman to come up with a fine sleek rainbow). As pre-arranged, Superintendent Blinn was hunkered down in the logjam just downstream with a twelve gauge loaded with two rounds of 7½'s, and three "double ought" just in case, ready to sting the haunches of Sister the minute she latched onto Rollie's fish. The show promised to be a good one, so a bunch of us who happened to be in the park that day sat up on the ridge above the river to watch.

Much to his credit, I gotta admit it didn't take Rollie long to hook a helluva fine trout. (Of course, we never did see it close enough to tell whether it was fair hooked or snagged.) At any rate, the fish started splashing like mad, and Rollie let out a few "whoops!" that I'm sure they could hear all the way back to his home in Wisconsin. When Sister didn't show up after the rainbow's first slap of the tail, Rollie eased off the drag so it could run out some line. Right on cue, halfway through the backing, that big familiar furry face jutted out of the willows on the far cutbank. Rollie set in to maneuvering the fish to just the right spot for his gunner, and Blinn eased off the safety and slipped a little lower among the logs. Sister just stood there sizing up the situation for a few minutes, trying to figure out how come Rollie was wearing that silly hat instead of his "Smokey" Stetson.

About that time, the trout made one more sprint back down toward Naknek Lake, and Sister made her move. She plunged off that six-foot bank onto that fish like a hungry hound going after a fumbled flapjack. Then quicker than the water could soak into her fur, she spun around, fish-in-mouth, and regained the dry shore before Blinn could draw a bead on her butt. Last we saw of Sister, she was making a blurred streak through the alders, while Rollie held his rod high and burned the shit out of his fingers trying to slow the reel enough to snap the leader. If I've ever laughed harder, I don't recall the time.

That evening we sat around reliving the details of the moment and ribbing poor Rollie right out of his boots. He got asked about the weight of the leader. Did he really think he coulda turned her? What the hell kind of net would he have landed her with? Would he have released her after the fight or tried to stuff her into the creel?

Rollie didn't say too much, though. He just sat back sipping his tea and grinning from way down inside. After all, he was in a unique class. Probably the only man ever to tackle an eight-hundred-pound bear on fly gear and be glad to tell about "the one that got away."

27

The Gift Eagle

DAN MULHERN

Shortly before three in the morning the wake-up call to my motel room interrupted a brief night's sleep. I was in Pratt, Kansas, with Vernon Tabor, another U.S. Fish and Wildlife Service biologist, to try to capture Bald Eagles at a relatively new nest site in rural Stafford County, just north of Pratt in south-central Kansas.

During the twenty-five-minute drive north to where the nest was located, I had plenty of time to anticipate what the day might bring. This nesting pair of eagles was here for its fourth try, having first arrived in 1994. Each previous year the nest had been destroyed by bad weather during egg incubation, but the pair kept coming back. This year they had successfully hatched two eggs. We waited until about the time we expected the chicks to take their first flight, and then we set out on our trapping expedition.

We had been monitoring nesting Bald Eagles in Kansas ever since the historic first nest was discovered in 1989 at Clinton Reservoir near Lawrence in northeast Kansas. Mike Lockhart, a professional eagle biologist from our agency, had live-trapped eagles for several years in order to color-band them. He trained me and helped me get my federal permit to live-trap and band Bald Eagles. With the help of climbers we now reach several nests each year to get at pre-fledged chicks. In this way we're able to measure and band all chicks in a nest at one time, without the time-consuming effort of setting and watching baited traps. As a result, I had not attempted any live-trapping during the past four years, and truth be told, I had yet to make a successful capture of a Bald Eagle myself. But the Stafford County nest tree could not be climbed safely; we had to resort to trapping. So that morning, hope and apprehension battled in my mind.

Vernon dropped me off on the blacktop at the bridge over the North Fork Ninnescah River. Despite the predawn darkness, I knew the eagle nest was about a quarter mile directly west of me, upstream. From our scouting the day before, I also knew that the riverbank to my east offered a few spots on which to make a trap set. We had gotten permission to access this trap location from Brad Johnson, a local farmer who owned the land. He had told us that the eagles liked to perch on the bank of the river to watch for fish.

We had also decided to try something a bit different here: setting a trap out in the pasture itself. It had been an extremely wet June, and there were numerous flooded areas in the prairie grass about four to eight inches deep—perfect for placing a large carp for bait. The problem was that I couldn't hide the padded traps under a light covering of soil or silt as I was accustomed to doing; instead, I had to nestle them into the flooded grass. When I finished, the traps were visually very obvious. Such visibility was not good, but I had no choice. This drawback was the first strike against us, but it certainly wasn't the last.

There's a kind of excitement about being out in the dark, staking baits, and making trap sets in hopes of capturing a live eagle. You have to position the baits and the traps so they look as natural as possible and then "wound" the bait to make it look as if another bird of prey has been feeding on it. All this has to be set in such a way to fool a predator with eyesight twenty times better than our own. I'd come close, capturing a crow and a turkey vulture on separate occasions. But so far the eagles had always won, and that fact only added to my anticipation as I swatted mosquitoes and gnats with hands covered in fish slime.

I went back to the bridge, and Vernon soon returned after cruising back roads to see what our prospects were of finding a cottontail or jackrabbit to use as bait in an upland set. One of the hard-and-fast rules of eagle trapping is to use baits that are natural to the habitat. That means fish along water's edge and rabbits in the crop fields and pastures. As it turned out, we weren't able to bag a single rabbit on this trip, which severely limited our trap placement locations. Strike two, and the day was barely beginning.

We waited until first light so we could visually monitor the eagle family. Because of the lay of the land and the locations of the trees, we could see neither the eagle nest site nor the bait sites without driving within less than a quarter mile of either—too close, in our opinion, to watch without disturbing the eagles' activity. So we had the added frustration of sitting over a half mile away, waiting for a bird to fly high enough that we could tell if anything was happening.

Every so often we would drive up or down the blacktop road that split our two observation areas, the eagle nest to the west and the traps to the east. We dared not stop on the bridge too often in case we disrupted the eagles as they were heading toward the baits. This logistical problem turned out to be strike three, but we knew we had to make the best of it. The question was, how many more strikes could we sustain and still stay in this game?

People often ask what it's like to be a wildlife biologist. They want to know about all the exciting activities we get to do. Especially with a project as glamorous-sounding as "Bald Eagle trapping," folks imagine all sorts of things, evoking images of Jim Fowler scaling cliffs in remote landscapes. Eagle trapping, midwestern style, is a bit different. On this day Vernon and I would sit and watch an extremely sedentary family of Bald Eagles doing literally nothing for more than twelve hours. We eventually located both adults and one juvenile and kept track of them on and off during the day. We never found the other juvenile and still wonder about its fate.

During this time none of the birds ventured more than a few hundred yards away from the nest tree, content to make short circular flights every couple of hours. Days like this can be mind-numbingly boring, but once traps are set and baited, you have to keep an eye on them. You don't want an eagle sitting in a trap for hours in the hot sun. Brad Johnson came by a time or two during the day to see how things were going. He seemed genuinely interested in what we were doing and, I think, a little amused at our predicament of spending all that time sitting and waiting.

Actually, there was one bit of excitement about midday. During one of our drives to do a visual check we noticed that something didn't look right near the stream bank bait set. So we stopped to look at it with binoculars and saw a couple of turtles moving around on the site, and

I could see that at least one of the traps was sprung. As I looked further, I realized what had happened.

"Well, Vernon, add another species to our trap list," I said dejectedly. "We just caught a turtle."

I walked out to the trap and freed a big Red-eared Slider. His buddies had deserted him on my approach, leaving only a couple of scales remaining of the large fish I had put there for bait. Both traps were sprung, one on the hind foot of this frightened, unharmed, well-fed turtle.

As I carried the empty traps and bait stakes back to the truck, I thought about how I was clearly violating another cardinal rule of eagle trapping. Never, ever approach the bait sites in daylight; otherwise, the eagles may see the action and associate the baits with human activity. Here we were, already up to strike four, and day one was only about half over. I was feeling less than optimistic about the prospect of capturing an eagle.

Around five-fifty, just about the time we were thoroughly convinced the day was a washout, one of the adults did the unthinkable and actually flew away from the nest site toward the trap set! We watched in anticipation as it crossed the blacktop, eventually losing altitude to come down in the pasture, seemingly near where our remaining carp and traps were located. It was all we could do to wait a good fifteen minutes before driving down the road to see what was happening. The bird was indeed on the ground, probably not more than twenty yards from where we thought the baited set was. Clearly, the eagle could see the fish and was thinking about making a move.

To avoid frightening the quarry now that we were this close to success, we continued up the road until we were out of sight. Now began a series of waiting, driving past, then waiting some more. The bird never seemed to move its position. Finally, after forty-five minutes or so, we saw the bird in the air again, flying back toward the nest tree. Helplessly, we watched it disappear into the grove of trees where we knew its mate and progeny were patiently waiting. A whole day of watching nothing, and then this? I walked out to the site and picked up the untouched carp and the equipment, and we drove back to town for the night.

"Well, at least he's seen the bait now, and we haven't seen them eat all day," Vernon said. "Maybe tomorrow will be the trick."

We were trying to encourage each other, but I don't think either of us felt too enthused.

Next morning I staggered from the bed on getting the wake-up call, only to open the door in the face of a massive thunderstorm. *Great*, I thought. *Just what we needed this morning.* All we could do was keep an eye on the weather until it settled into a light drizzle shortly after five. I could see that the sky, which was already lightening from the impending sunrise, was clearing. It was around six o'clock by the time we arrived at the bridge, well after daylight and too late to be making eagle sets.

As we drove up to the bridge, we were surprised to see not one but both adults flying from east to west across the road. Maybe they had thought about that big carp all night and had been waiting for first light to find it again. How would they react when they found the pasture devoid of food? Would this be their only attempt of the day? We had little more than half a day to try, and I wondered if we'd already missed our shot. We debated whether it was even worth making the sets. Finally, I decided that this was our last chance, so we had nothing to lose. I walked out to make the same two sets I'd made the day before. Vernon stayed on the bridge this time in hopes of intentionally discouraging the birds from flying our way until I was finished.

When I got back to the bridge, Vernon had bad news. Both birds had taken flight and ventured in our direction. Though they hadn't come all the way across the road, they were high and close enough to easily see me out in the pasture placing baits. I was too discouraged to even try to count which strike this was against us. There was nothing to do but press on and play out our hand. Thus began day two of waiting out of sight of both baits and birds, driving past every now and then to investigate.

As it turned out, the eagles were in a different mood that day. It was sometime shortly before nine o'clock when the first adult decided that the carp in the pasture was just too enticing to resist, and the second adult followed shortly thereafter. At first they flew off to the east, well past our baits, and appeared to be hunting along the river channel quite a ways downstream. But it wasn't long before they both ventured back nearer the bait sites, settling down into the pasture. Again, we made a couple of drives back and forth to see what was happening.

Finally, we spotted what we had been hoping for—both birds were standing on the riverbank within five feet of the bait set! But they didn't move in on the bait; they just kept watching it. So we drove to a point where we could see what was happening, although it was closer than I would have liked. I saw movement in the water and looked through my binoculars in frustration at a couple of turtles dismantling the fish, virtually at the feet of the eagles. We drove out of sight, and then, as we turned to get a better look, we saw both birds flying quickly back toward the nest, and one appeared to be carrying something.

We checked the set and found no turtles caught this time, but one of the two traps was sprung, and there was only half a fish left. Did the turtles spring the trap, frightening the eagles? Did the turtles get half the carp, or did the eagle carry it away? If the eagles took the fish, would it satisfy their hunger, leading to another day of inactivity?

The carp and traps we'd set in the flooded grass, away from the channel, were still intact, but I didn't feel very confident that the eagles would visit this location. As we carried the remains of the disrupted bank set back to the truck with us, that old feeling that *this was just not meant to be* was settling into my gut. We went back to our vigil, neither of us wanting to voice our pessimistic premonitions.

The thing about working with wildlife is that you never know what's going to happen next. I don't know if those eagles were naive, curious, or just plain hungry for an easy meal, but they didn't stay away long. Sometime around ten-thirty both birds flew back to the pasture. One adult flew way downstream and settled in to stare at the Ninnescah River flowing past its low perch. But the other soon moved into a position very near our carp. Each time we drove past, it seemed the bird was in a slightly different location, seemingly getting nearer the bait, until we decided it must be virtually right on top of it. But for over an hour, each time we made a visual inspection, both birds maintained their positions, the nearer bird giving no outward indication it had encountered a trap.

This bird remained standing the whole time, although a couple of times we thought it might be standing in an odd posture. Brad had come back again this morning, and he was with us as we were discussing strategy between drives. When he sensed that something could be happening, he went home to get his family, and we had the added pressure of trying to come through for an audience.

Eventually, we could stand it no longer. Rather than drive all the way past and out of sight, we parked the truck as far away as we could and still see the bird with our binoculars. I put the spotting scope on the bird to get a better idea of what it was doing. A couple of times it put its bill down near its feet. Feeding? Trying to free itself? The uncertainty was frustrating. Finally, it happened. I saw the bird flap its wings to attempt to get airborne, only to be held to earth by an unseen force. It was clearly caught!

We raced back down the road, screeching to a halt near the bridge. With my hip waders on, I cleared the barbed-wire fence in record time and ran as fast as I could out toward the bird. The eagle, meanwhile, had seen our arrival and was in no mood for a close encounter with a puffing biologist with heavy rubber feet. It took off across the pasture, trailing five pounds of trap and drag chain behind it.

The eagle headed straight for the river channel, exactly where I had hoped it wouldn't go. I knew the water was not deep, but I also knew the middle of the channel was over the head of a Bald Eagle with a heavy chain on its foot. A drowned bird was not the conclusion I had envisioned in my dreams of this moment. The bird hit the channel long before I caught up with it, flailing water high into the air with the beating of its wings.

Fortunately, eagles are remarkably buoyant, and those long wings served as oars to propel it across the river. I hit the water as the bird made it to mid-channel, and we both reached the far bank about the same time, a distance of no more than twenty-five feet in this narrow drainage. The trap chain had dragged along the bottom but had not pulled the bird underwater. Although I took lots of water into my waders, I was able to catch the bird carefully by one wing, making sure it didn't keep going farther downstream.

At that point I sat down on the bank to catch my breath and calm down from the adrenaline rush that had propelled me during the chase. I was holding onto one wing of the eagle, which now floated on its back, trying to defend itself with the long talons of its one free foot. I felt underwater for the trapped foot and quickly grabbed both feet in one hand. Then I wrapped my other arm around the eagle's wings to protect them, cradling the bird to my chest as I slogged back across the channel toward Vernon, who was waiting on the far bank.

He helped me free the bird's foot from the trap, and it was about this time that we realized we had finally accomplished our goal. We had successfully (and without injury) trapped a free-flying wild Bald Eagle! Brad and his family were there, taking photos and helping us get our equipment rounded up, and they eventually assisted with the measurements and recording of data.

After taking measurements of various external features such as bill and talons, we identified the eagle as the female of the nesting pair. Although the release of this waterlogged bird was hardly majestic, she managed to get herself airborne and back to the nest. She was none the worse for her experience and now carried an identification band on each leg. One was a plain aluminum band with small characters, and the other was a bright purple band reading "2/Y" in large characters that would be visible from long distances with binoculars.

She was the twenty-ninth eagle banded since 1989 in Kansas, part of an ongoing marking and monitoring program. But for Vernon and me, she was bird number one—the first bird successfully live-trapped without the help of my friend and mentor, Mike Lockhart (who would have cringed at some of the things that happened to us in our attempts to get her). That's the funny thing about this whole affair—so many things went wrong that by all rights, we should not have even gotten close to capturing an eagle. When it comes right down to it, we really didn't deserve that bird. But sometimes the good Lord overlooks these things and grants our wishes anyway. That's why I'll always remember bird "2/Y" as my gift eagle.

28

Tequila

PATRICK LOAFMAN

We stand beside Highway 101, on the corner by the marsh where the Dosewallips River empties itself into Puget Sound. A blue heron stands motionless in the shallow water. Across the still waters of Puget Sound, Seattle, Tacoma, and Olympia buzz like electric beehives.

Two pickup trucks with Washington Department of Fish and Wildlife logos on their doors turn off the highway onto the gravel road, kicking up dust. The heron lifts into flight with a guttural complaint. A young man with a flat-topped haircut hops out of his truck and approaches us. "Hi, I'm Matt," he says, holding out a hand.

"I'm Pat." We shake. "This is my wife, Kim."

A chopper flies overhead. A man with mirror shades stands by his truck, parked next to Matt's. He has a badge and a pistol in a holster. A redheaded woman steps out of Matt's truck and introduces herself to Kim, then me. I instantly forget everybody's name.

The radio calls out numbers.

"This is wildlife 230," the man with the bristly flattop says into his radio.

"We're going to fly along the top of the ridge and see if we can locate them there," the radio voice says. We can hear the chopper in the background.

"Ten-four. We'll drive the road down here and see if we spot them. They've been seen in that big clearcut on the west side of that ridge recently," Bristlehead answers. "You two stay here. We're driving up there to see if the elk are in this patch of woods," he says, pointing to a distant ridge, "but we'll probably be back by here soon." He hands me a radio from his truck. "I'll leave this with you guys so you can hear what's going on."

He and the other two hop into their trucks and speed off.

A few Song Sparrows call from the nearby bushes. A kingfisher hovers above the water, and then it dives and comes up with a fish and flies away. It's one of those rare sunny days in December, making it seem like spring.

"We've located the herd," the radio calls out. "They're at the top of the ridge next to that green trailer."

"Go ahead and shoot one of them," Bristlehead responds.

"Ten-four."

The two trucks come driving back. We follow them down the highway and turn right onto a dirt road, winding through clearcuts and young Douglas fir plantations. At the top of a ridge we all park at a pullout with a bird's-eye view of Puget Sound. Downslope is a herd of forty elk, staring up at us and at the downed cow with a dart in her hip. There are ten cows, a few calves, and a few bulls. The tranquilized cow is lying on her side. Sleepy-eyed, she looks up at us but cannot get up.

We all get out of our trucks. Kim and I stand next to the redheaded woman.

These elk in the Dosewallips watershed migrate into Olympic National Park's high country in summer, but in winter they are down near Puget Sound, sometimes coming into people's backyards to nibble on costly landscaping and fertilized gardens. They know this drainage better than any human can, migrating the full length of it each year. I can imagine how important the older elk must be in leading the herd to known foraging areas. The alpha cow, an old female, is the leader of the herd.

"Have you ever done this before?" whispers the redheaded woman.

"Nope," I answer. "We've been radio-tracking the Sequim elk herd this fall, but we've never handled a downed elk like this."

I have handled many animals as a wildlife biologist—everything from amphibians, reptiles, songbirds, mice, rats, and spotted owls to flying squirrels and even alligators and sea turtles—but this is the first time for elk.

The other two men come over, and we stand in a huddle.

"She's still conscious; she can hear us talking and moving about,"

Bristlehead whispers. He looks back at the elk and then continues, "We'll sneak up quietly behind her; you all follow us."

We follow Bristlehead in a line off the road over the scattered wood of the clearcut. "Watch her feet. A kick can kill you," he whispers along the way.

The elk looks over her shoulder as we approach, and she manages to get up and stumble downslope toward the rest of the herd.

"Damn," says Bristlehead. "Let's back off."

We walk back up to the trucks. The elk stumbles about a hundred yards and then lies back down. Bristlehead discusses the situation with someone on the radio. "The old poison would have knocked her right out, but this new stuff is too weak," he says to the warden.

"But it was more dangerous, wasn't it?" the warden asks.

"Yeah."

A few silent minutes pass as we watch the elk lying there.

"We're going to have to sneak up and jump on top of her. I'll give her another injection when we've got her under control."

We walk behind Bristlehead and the game warden again. They sneak up and jump on the elk. One of them puts a hood over her head; the other grabs her hind legs and ties them together. We lie on the elk's body and try to hold this eight-hundred-pound animal down.

"We have to flip her! She's in a bad position; it'll make it hard for her to breathe."

We all get on one side of her and heave. She flips over to her other side.

Bristlehead gives the elk more injections. She begins breathing slowly and deeply, snoring with each breath.

"Hold the head in an upright position, Kim—we gotta keep the airway open."

"Here, shove this up its ass and get a temperature reading," Bristlehead says, handing me a thermometer.

I take the thermometer, lift the tail, and examine the situation, then let the tail fall down again and take a few breaths to get up my nerve.

Kim laughs. "I'm glad I don't have your job!"

I lift the tail and insert the thermometer. The elk continues her deep snoring breaths with the metrical rhythm of a slow clock. I begin breathing with the rhythm unconsciously.

"You've pulled teeth before, haven't ya?" Bristlehead asks the warden.

"I've pulled cougar teeth."

"That's close enough. Take this and pull the whistler," he says, handing the warden some pliers. "You're gonna have to wiggle the tooth back and forth to loosen it."

"One hundred and eighty degrees," I say, still holding the thermometer.

The elk's snoring is steady and sure, as if pacing our work.

"Her teats are swollen; she's nursing a calf," Bristlehead comments, then turns to me. "Here," he says, handing me a rubber glove and a tube of KY jelly. "You can collect the fecal sample. Reach up there and get a big ol' handful." He says it straightforwardly, but I bet he's laughing to himself.

I look up and see Kim prying open the elk's mouth and the warden reaching in with the pliers.

I put on the glove, lift the tail, and wonder to myself if this is physically possible. It is not a large opening, but I discover it is possible.

"Fecal sample collected," I say somewhat proudly. "Where do I put it?"

Bristlehead hands me a bag.

"Got it!" The warden stands up straight, holding a bloody tooth in the pair of pliers. "I oughta been a dentist. Hey, Matt, didn't you say you had a tooth bugging you?"

Bristlehead laughs. "Clear away all this gear," he says as he jabs a needle into the elk's leg and begins drawing blood. "Soon as I get this blood sample, I'm gonna inject the antiserum, and when I do that the elk is gonna jump right up. We've had them run over all our gear before."

The elk continues snoring. If only she knew all that was going on while she was sleeping. In addition to having blood drawn, a fecal sample dug out, a tooth yanked, and getting poked half a dozen times with needles, she has been fitted with a radio-transmitter neck collar so she can be tracked. Her tongue is hanging limply out of the side of her mouth, and blood is dripping out.

Bristlehead and Warden untie the legs. We all back off. The elk's head lies on the ground, its tongue in the dirt. Then Bristlehead injects the antiserum. "Damn," he shouts halfway through.

He joins us. "The needle came out of the jugular as I injected," he says. "I didn't get it all in, but I think it was enough."

"Its head is lying downslope," the warden observes.

She should be jumping right up.

Silence follows. No more snoring.

"Did it quit breathing?" Warden asks. With his mirrored shades off, I notice his eyes are a soft blue.

Silence. The regular clock-like breathing has stopped. We all stand there silently as seconds stall. I remember other times in the field when captures didn't go as planned. I remember a sparrow that died in my hands as I was banding it. It went cold in less than a minute. Then there was the flying squirrel that went into shock and died as I held it. Although these unfortunate accidents are rare, the thought of them sends a shiver up my neck.

Kim stares at the elk; the redheaded woman looks down at her feet; the warden shifts his weight from one leg to the other; Bristlehead looks up as if in prayer.

Is the elk dead?

The scene is silent. Not even a common Song Sparrow can be heard rustling in the bushes. And all the elk look up at the downed cow. If the elk is dead, what do we do? Just pack up our gear and drive away? And would the herd come up to her and smell her, nudge her limp body, the way elephants return to bones of their dead relatives, reminiscing?

Bristlehead begins rummaging through the gear, breaking the silence. He runs down to the elk, straightens her head upright, and jams in another needle. The elk gulps in a breath of air, and Bristlehead runs back to us.

The elk's regular snoring returns, but her tongue still hangs limply out. Slowly she opens her eyes and looks around.

"Thatta girl."

We all breathe deeply.

She looks up at us, shakes her head, and then stumbles uneasily to her feet.

"Hell of a hangover," the warden jokingly narrates, putting his sunglasses back on. "Must have been that tequila. Tequila will do it to you every time."

29

Polar Summer

CAMERON WALKER

Sonia holds a silver funnel over her nose while I strap it around her head. In the corner Donie peels red labeling tape off a roll and sticks it along the sides of her XtraTuffs, the insulated, waterproof boots we wear in the boggy tundra. The brown boots slowly transform into ruby slippers. Kristen cuts a lion's mane out of a grocery bag while Laurie wraps tin foil around her arms and legs. This is how the Fourth of July is celebrated hundreds of miles north of the Arctic Circle.

Our five-woman team is working at a research station in northern Alaska for the summer. Donie is the postdoc who is leading our project, and the rest of us are undergraduate field assistants.

The research station consists of seven trailers, a row of tents, and three outhouses, which are covered with graffiti only a scientist would understand. It all sits on the shore of Toolik Lake, and the Brooks Range glows in the midnight sun to the south.

Toolik means "loon" in Inupiak, and I can hear the loons calling from the inlet at night. Maybe they laugh because they think they have driven the darkness away.

Returning workers have been telling us about the Fourth of July festivities since we got here in June. It's the one day of the summer when no one works, when the beakers and test tubes are left with their magnetic stirrers turned on and the LI-CORs record data on their own. Dinner will be a feast. All the big shots will be there, the ones who are listed in the indexes of *Nature* and *Science*.

Our role, as the peons of the scientific community, is to provide the entertainment: a Fourth of July parade, complete with floats and costumes crafted of the best Toolik has to offer. Tarps, duct tape, Carhartts, and PVC will become architectural marvels. My little group has chosen a Wizard of Moss theme, a tribute to the plants we've been examining all summer.

I'm sitting on the counter in our lab trailer, stuffing moss into my overalls and underneath my cuffs. Watching my four colleagues bump elbows as they struggle to arrange themselves, I think about how far we've come. We've been at Toolik for a month, learning to live with grizzlies, swarms of mosquitoes, and thunderstorms that roll across the tundra. I'm relieved that, so far, working together hasn't been as painful as our first days together.

The day we left Fairbanks, in early June, we got on the road a little later than planned. Everything seemed to take a little longer with this group. Laurie, Kristen, and I were always on time, and then we were stuck waiting for Donie and Sonia. We had packed our blue monster of a Suburban the night before, asking Donie and Sonia if they were sure they had everything they needed for morning. "Oh, yes, yes," they replied. "We're fine!" The next morning they were rummaging through the back, in search of an unmatched boot and a tampon.

The Suburban has two huge cracks snaking across the windshield from previous trips up the Dalton Highway. The highway rolls alongside the Trans-Alaska Pipeline, which carries oil to Valdez from Prudhoe Bay.

The oil companies close Prudhoe Bay to the public. The good folks in the RVs and rented minivans must stop in Deadhorse, a few miles from the Arctic Ocean. I watched the tourists passing by, going the other direction, and wondered why they had driven so far. We weren't even going as far as Deadhorse, and already I was questioning why I had come.

I was sitting in the middle of the backseat, and on a few hours' sleep, I was already cranky. On my left, Sonia picked invisible dirt off her stiff new Carhartts. For the love of God, I thought, they're work pants! They're supposed to be dirty! Throughout the drive she peppered us with questions about tents, backpacks, and raingear, using the precise pronunciation and specific word choice of someone who didn't grow up speaking English. She seemed the least prepared for the adventure ahead.

Donie, the postdoc, was seated on my right. Her graying hair was tied in its usual braid, and through her thick, round glasses she was studying *The Geological Formations of the Dalton Highway*. She looked up only to check the numbers on the mileposts bouncing by to ensure that she was reading from the appropriate section. She droned on about alluvial plains and culvert ponds, unaware that no one was listening.

The highway is one and a half lanes of gravel, with soft shoulders that slope down to the tundra. Trucks come barreling down the grades. They know they will win any contest—with a bear, a moose, or a Suburban. Cars drive the road with their lights on. I wasn't sure, in this land of perpetual daylight, how that was supposed to help.

Donie was specific in her instructions to us before we drove. We were to leave the lights on at all times and pull to the right and stop when a truck approached us. The gravel is firmer in the center of the road, so she also ordered us to stay to the middle and off the shoulders.

We ended up rotating drivers every two hours or so. Kristen drove aggressively, speeding up over hills, leaning into curves. This driving style matched what I'd seen so far of her personality; an hour after landing in Fairbanks she had challenged a postdoc from Australia about his pick in the NBA finals. She packed a lot of energy in her short, stocky frame: she walked quickly, talked quickly and loudly, and her gestures looked like those of a child imitating an adult.

Laurie drove the way I did, with a little caution and hesitation and a bit of speed to compensate. I couldn't tell about Laurie at first— couldn't tell if I was going to like her. I felt I had a clear understanding of what the others were like, so apparent that their images were there when I closed my eyes, burned in dark and light on my retinas. But tall, quiet Laurie didn't make an immediate impression. All I knew was that she and I were the only ones who wanted to sleep in our tents for the summer. It seemed that she was just listening, and waiting.

After a painful hour with Sonia in the driver's seat trying to navigate the rows of potholes, I was happy to be behind the wheel. I liked the challenge of the road. Although I couldn't peel my eyes from the gravel ahead of me to look at the mountains, I felt as if I knew their secrets through the wheels against the road.

It started to rain, one of those afternoon mountain storms that dumps until you are nearly drenched and then rises to the next ridge.

Donie leaned against the passenger window, eyes closed, mouth open. The three women in the back were quiet. I flicked on the wipers and waved to a trucker coming over the hill. He saluted me, and I started to grin. We were both in it for the long haul. Suddenly I was glad to be going north.

The Suburban ground up one side of the pass through the Brooks Range and sailed down the other side. I never touched the brakes.

I expected the landscape to be different from anything I'd seen before, but as we came down from the pass, I was reminded of the eastern Sierra, with its large, desert plains broken up by snowy peaks. Maybe this land was not so different after all.

For the last fifteen minutes of the drive, Donie signaled me to slow down at the crest of every hill. She couldn't remember where the lake turnoff was, but she knew it was at the top of a hill. After passing a hand-lettered sign that read "Toolik Lake—No Services," I turned left onto a bumpier road.

We parked in front of the trailer that would serve as our lab for the summer. Boxes of equipment, data loggers, mechanical pencils, and 118 bottles of beer were unloaded into the rickety green box.

The next morning we hiked up to our work site. A mile-long boardwalk had been built between camp and the site so we wouldn't disturb the sensitive tundra as we made our daily treks. Over time I learned where I could run ahead without looking and where to watch out for loose boards. I also discovered one board that was springy enough to launch an unsuspecting co-worker into the air.

As we neared the site, the boardwalk ended and we fanned out across the land. Walking through the tussocks, small mounds of earth and plants on the tundra, was a challenge. That first day, after we had learned by trial and error to walk around the mounds, Laurie said, "I feel as if I'm trying to walk across a crowd of people. Their heads are in the way, and I'm trying not to hurt them by standing on their shoulders."

I smiled. Donie nodded, squinting up at the sky, and Kristen charged ahead to the site.

The tundra ecosystem, with its unique plants adapted to the radical change in seasons, is thought to be the first ecosystem to register change due to global warming. Our work was to look at the effect the absence of one species, or group of species, would have on the rest of the tundra flora. Our site consisted of more than one hundred two-by-three-meter rectangular plots. From each plot we removed something—moss, deciduous plants, evergreens. We then used a grid, which we made out of plywood and twine, to estimate the species composition remaining in each plot. These measurements would be compared to the composition measured in subsequent years.

Although the cause may sound noble, the labor was not—we spent five weeks flat on our stomachs or crouched on boards, poking at bits of moss and lichens with names like Dead Man's Fingers.

Eight hours picking moss became a time to learn about one another. We heard the stories of how our parents met, how many brothers and sisters we had, what we had been like in high school.

I learned that Sonia had left Iran in order to make better use of her degree in German literature and that she had never been hiking before coming to Alaska. I was surprised by how adventurous she turned out to be. She went on all the hikes and never complained about the rain, the mosquitoes, or the fact that Donie and Kristen treated her as if she were in kindergarten. I think she was relieved when we returned to Fairbanks at the end of the summer, but I also know she started hiking on her own when she got back to school.

I learned that Kristen had just transferred to U.C. Berkeley from a college in Worcester, Massachusetts, and that she was lonely, although she never said it.

And as we moved from talk into comfortable silence, I found that Laurie was the one I got along with the best. Sometimes, in an afternoon rainstorm, Laurie and I would sing songs from old musicals while Kristen would calculate how fast she had to pick her plot in order to finish the row by dinnertime. Sonia and Donie, meanwhile, would still be working on the same plot they'd been working on all morning.

After putting the finishing touches on my scarecrow costume, I sit on a counter in the trailer, swinging my legs and laughing at Sonia in her Wicked Mosquito of the North outfit. When the last piece of duct tape is in place, we march through the screen door to await the parade.

The parade begins with a lone figure wearing a tarp, waving a dirty flag. Some of the truckers and transportation workers stop in after word spreads over CBs that a feast is in store. They hoot at the flag bearer and hand her a beer.

Following her is a raft perched atop the camp truck, the bow tipping down so it covers part of the windshield. Four of the professors sit inside the raft, wearing helmets and life jackets. Each takes an oar and begins to row while singing a twisted version of "Yankee Doodle Dandy."

We follow, singing "Somewhere Over the Brooks Range." At the end we dump a pail of water on Sonia, and she withers away like a proper wicked mosquito.

Behind us come researchers on tricycles, people dressed as fish, and a woman covered entirely—and only—in moss. We parade around the trailers and end up on the deck of the communications room, where the feast begins.

Someone has screened the deck with mosquito netting so we can eat outside overlooking the lake. Laurie, Kristen, Sonia, and I meet up with the group of guys we've been fly-fishing and hiking with all summer. We all cram around a tiny wooden table, knocking knees and sloshing bottles of prized Alaskan Amber Ale. We feast on fried chicken and gardenburgers topped with the works. For dessert, there are cherry, mocha, and chocolate cheesecakes, and ice cream bars covered with sprinkles. Everything vanishes within minutes.

We sit quietly for a long time after eating, looking out at the lake. The sun swings a little lower in the sky. People slowly drift away, off to find another stash of beer or to sleep off the fireworks exploding in their heads. Eventually, there are only four of us. Laurie's eyelids hang lower than usual. Andy and Jeff rest their scraggly chins in their hands.

"What should we do now?" someone asks.

After a moment, Laurie replies, "Sauna?"

We smile and push back from the table. "See you guys there."

The sauna perches on the edge of the lake, with a long dock that extends out over the chilly water. Bathing in the sauna is the only way to get clean up here. Trucking out the gray water from showers and sinks is too expensive. Three nights a week the women have the sauna for two hours, and then the men. Late at night the sauna is co-ed, but Laurie and I had never gone before.

At first, even sharing the sauna with other women made me uncomfortable. Most of the women here walk around in their skin as if it really belongs to them. In my previous locker room experiences I'd noticed a different look on people's faces—and certainly felt it on my own—a look that said, "This isn't my real body. That one is at the cleaners today." But each week the sauna became a little easier, a little more cleansing. Eventually, I began to worry more about the mosquitoes than the other bathers.

It is after midnight when Laurie and I walk down the path toward the sauna. A tiny sliver of the sun hides below the horizon, the lowest it will set. The Brooks Range brightens with pink and orange, aflame with the sun's rays.

We stuff the stove with wood. The arrow on the small thermometer climbs higher and higher, and we pour water over the coals and over ourselves. When we finally get too hot, we run out of the sauna, down the dock, and into the icy water. We laugh louder than the loons when the boys vault over us.

Gasping, I pull myself back up onto the dock. I fill a pitcher with water and pour it over my hair, leaning back as the water streams down my face. As I soap myself, I look out at the lake and the tundra and the mountains. The tundra is alive with the colors of autumn leaves. A light breeze keeps the mosquitoes from landing. I scoop a pitcher of hot water from the barrel by the stove and let it run down my back. I don't know if this is what you would call northern freedom, but it sure feels good.

30

JAMES LAZELL

In March 1980 I was sent by the Nature Conservancy to the British Virgin Islands to work for what was then the Department of Natural Resources and the Environment. I set up at the Seaview Hotel in Road Town, Tortola, and began island surveys. The hotel owner, Ishma Christopher, was widely renowned for her supernatural powers. She also made the hottest, finest, and tastiest hot pepper sauce that I, regarded as a hot sauce connoisseur, ever savored. She kept me out of trouble with sage advice and made sure I met all the right people. She was the first of many to tell me of a wonderful, unspoiled, largely forested island called Guana. "It's too bad you can't go there," she said. "It is a really great place, and a natural habitat." After hearing several such statements from other people, I finally asked why I could not go there. "Oh, it's a privately owned island. They won't let anybody on it." Well, there were at least forty-seven other islands to inventory, so I was a busy man. I did not worry much about Guana Island. One day, however, I was looking something up in the telephone book and noticed a listing for Guana. I dialed the number, and island manager Mary Randall answered the phone. "I'm down here doing biological surveys of the islands..." I told her.

"Come right over!" Mary said. "I'll send the boat to pick you up at the dock on Beef Island. We would *love* to have a biological survey."

The initial survey I conducted indicated that Guana is a most remarkable island. It is privately owned, but it is also a destination resort hotel—at least for ten months each year. But nowadays, for the other two months it is turned over to biologists. More than twenty years ago I convinced the owners that Guana could be the site of major and important research in ecology, biogeography, and systematics. I also got island management to agree to stock Auntie Ishma's hot pepper sauce. From the beginning, I set out to bring colleagues from all over the world to work and study on Guana Island.

When Liao Wei-ping first came from Guangdong Province, China, to Guana to study birds in 1984, the Brown Booby was the only species of bird known to be present that also occurred on his native Hainan Island. He yearned for a museum specimen to take home to his base, the Guangdong Institute of Entomology's Department of Zoology, Academia Sinica. I fixed him up with a .22 rifle, but even if he'd had the chance to shoot at a booby, the bird would most likely have flown irretrievably out over the sea. Liao is not a strong swimmer, and anyway, the best opportunities to shoot a booby were from cliffs no one would willingly jump off. Days went by with no success.

Coincidentally, Liao was also trying to catch a Scaly-naped Pigeon in his mist net, set across the top of the driveway parking area in the col between Anegada and Barbados Houses, two of Guana's guest residences. The pigeons regularly rocketed through this col but always over Liao's net. Anthropologist Michael Gibbons was in residence, digging subfossil bones and Amerindian potsherds. He walked by one day and observed this phenomenon and Liao's despair. "I'll get you one," said Gibbons. "You just have to call them down: *Coo-ee, cooo-eee...*" Swoosh! A fine pigeon hit the net, bagged instantly. Gibbons is not without a sense of humor: "Have some Auntie Ishma's hot sauce."

But Liao's jubilance was short-lived in the face of his failure to get a booby. I knew of a rock close inshore, but just out of sight from the beach, where a booby usually perched each morning. Gibbons and I hatched a plot.

We told Liao to stay in the temporary lab we annually set up in the common room of Anegada House and work on his notes and specimens. Gibbons, we said, would call in a booby and shoot it—in flight, of course—so that it would fall right at the Anegada House door. Liao chuckled and agreed; he could hold up his end of the deal if Gibbons could manage his. Early next morning Gibbons and I set off for the chosen rock, which was plenty far away from Anegada House to be out of audible range. Sure enough—there perched a fine adult booby. I am by no means a crack shot; only by incredible luck could I have hit a flying booby with a rifle. However, this was a job I believed even I could accomplish.

I lay prone on the ground, rested the rifle solidly on a big rock, lined up the booby in the buckhorn sights, and gently squeezed off a round. The booby toppled into the shallow water, and we waded over and picked it up. Back up the hill we went; Gibbons took up his station just outside the south window of the common room, just inside which Liao was working. He gave me a minute or two to lay the dead booby on the doormat and take up my station below the steps leading up to the main door.

"*Booo-bee, booo-bee, booo-bee,*" called Gibbons. Bang! He fired the .22 into the air, and I slung my leather water flask over the porch, banging it into the door and bouncing it out of sight into the bushes. Liao came running, yanked open the door, and beheld his prize, incredulous. He had his bird for direct comparison with his Chinese specimens back home.

But that is not the end of the story.

Liao immediately set about preparing his specimen. When it was skinned—carcass bared, skin wrong side out—Liao called us over. His increased incredulity was written all over his face.

All three of us carefully examined the skin and carcass of that booby. There was no sign of a bullet hole. I never said I was a good shot, did I? On the other hand, I never go anywhere without a bottle of Auntie Ishma's hot sauce. I do not want to lose my shooting edge.

Liao prepared a fine representative collection of voucher specimens from Guana Island. It is now at the Guangdong Institute, Academia Sinica, and remains a permanent testimony to his diligent efforts—and one small mystery.

31

Sandhill Season

JENNIFER BOVÉ

Oh God, not again.

I shoved the door of the truck open, leaned out far enough to miss my boots, and threw up. It wasn't a surprise, but it wasn't getting any easier either.

Such bouts of nausea had become a regular part of my schedule in the four months I'd been pregnant, as if my body was hell-bent on starvation. I'd never been one to skimp on food during long days in the field, and now, when I surely needed a little extra fuel, a plum and a couple of Saltines were purged from my system like poison.

Swiping my hand across my chin, I stepped over my breakfast as it soaked into the dirt. Hungry or not, there was work to be done. I hauled my chest waders from the bed of the F-150 and leaned against the wheel well to tug them on. Under the cover of the waders I unbuttoned my jeans to allow a bit of comfort room. I hadn't yet conceded the need to wear the stretch-bellied maternity pants my sister had given me. I wasn't quite that far along, mentally speaking anyway.

With a backpack full of data sheets and my field notebook, hand-held GPS unit, measuring tape, binoculars, radio, water bottle, and granola bars that I wouldn't even attempt to eat, I began to plod down the hardened bank of Bird Creek through Conboy Lake National Wildlife Refuge.

It was June, and I was looking for a nest that had been built, and presumably abandoned, by a pair of Greater Sandhill Cranes. I'd been observing Conboy's small population of cranes (the only known breeding population in Washington state) with a spotting scope on valley overlooks since February. I had identified twenty-five individual cranes, many by the brightly colored identification bands fitted on their legs by refuge staff in years past, and I'd mapped the nesting territories of nine breeding pairs. With my eye pressed to the scope for hours at a time, I had monitored the cranes' behavior, documenting their territorial scuffles, mating dances, and nesting activities.

A telltale sign that eggs are present in a nest is the "nest exchange," a changing of the guard between Sandhill parents. One crane leaves the nest to forage for a few hours while the other tends the eggs, and then they switch. The pair whose nest I was seeking, known around the refuge as the "C&H" pair, had engaged in regular nest exchanges for the past few weeks. Then I'd observed an abrupt change in activity: both cranes had left the nest to forage in an adjacent landowner's pasture with no hatchling colts in tow. It appeared as if something had gone wrong at the nest, and it was my job to try to figure out what had caused the cranes to evacuate.

I liked being on the ground inside the refuge. So often the distant observer spying on the furtive rituals of coyote, elk, and crane, I was excited to move along the very paths the animals tread. As I approached the C&H wetland, though, I could see that finding the nest was going to be a little tougher than I'd expected. Pinpointing a nest on aerial photographs did not present anything near the challenge of finding one on foot in a soggy, matted, and life-sized wetland. Without the advantage of the bird's-eye perspective I was used to, I was going to have to find my way over land (and through water) on equal parts inference and luck.

A downed tree spanning the deep trench of the creek was the only bridge by which I'd be able to reach the C&H wetland, and standing beside it I didn't feel too confident about hopping across it the way I might have done four months ago. So I got down on my hands and knees and sort of schlumped across with a lack of grace unique to a pregnant woman in waders, and once I'd made it to the other side, I could see the wetland glistening through lush clumps of *Juncus*, as still and silent as a secret waiting to be told. The cranes' nest was out there, somewhere.

Since I didn't have any idea how deep the water in the marsh might be, I grabbed a stout willow and felt my way cautiously down through soft mud until the sole of my boot met firm footing about two feet below the water's surface. Not so bad. The water grew deeper, though, as I slogged away from the bank. I couldn't help feeling a little nervous when the water crept toward the top of my waders, sealing the cool neoprene tight against my body. And the ground was becoming more difficult to traverse as well. There were mounds of decaying vegetation and deeply pocked elk trails everywhere I stepped, and I just prayed that I wouldn't stumble onto some large drowned beast in the murky water.

But I made it into the middle of the wetland without incident, holding my backpack up around my neck to keep thousands of dollars' worth of federal contents dry, and I easily spotted what looked like an unbound bale of hay lying amid the rushes: *the nest*. My heartbeat quickened as I inched toward it.

The nest was a perfectly dry mat, nearly five feet in diameter, built of reeds and grass in about three feet of water—a floating island in the apparent safety of the heart of the marsh. No coyote or dog would bother swimming so far to prey on a crane's nest, risking the blow of a sharp and lightning-quick bill to the skull. If a predator had indeed disturbed this nest, it would more likely have been semiaquatic or avian, and the clues I needed would lie in any egg fragments that remained.

At first the nest looked empty and untouched, but when I stretched to peer around the side of it, I discovered pieces of brown speckled eggshell wedged among the nest fibers at the edge of the water. I gently plucked them free and laid them out on the nest in front of me. Most were small chips that would offer no hint of the marauder's identity. One, however, was nearly intact, still smeared with blood and bearing the distinct marks of two canine teeth. Just what I'd hoped to find.

I slipped my backpack from my shoulders and placed it on top of the nest so I could pull out my measuring tape and field notebook. The size of the canine punctures and the distance between them revealed the identity of the predator as river otter, a rarely seen inhabitant of Conboy Lake but one that could potentially wreak havoc on the crane population.

Methodically, I made my notes, marking down a GPS point that would allow precise mapping of the nest for future reference. I also measured the nest and the water depth below it. These characteristics would further our knowledge of the cranes' nesting behavior and help shape habitat management guidelines for the refuge. My morning's efforts had reaped successful data.

Satisfied, I placed the eggshell in a plastic bag to take back to Headquarters, and then I heard the cranes.

Tinny, resonant voices rose in plaintive rounds from beyond the willows, so near that I bet they could see me even though I couldn't see them. No doubt it was the C&H pair, wondering why a human had come to pick through the ruins of their nest. I imagined them roaming the lakebed dispossessed, and I felt guilty for quantifying their loss on the pages of my notebook.

When I thought about it, those broken eggs meant a whole lot more than good data. There had been babies inside them, warm and waiting, not unlike the one I carried in me. Tears stung my eyes, and at the same time, I wanted to boot myself in the ass for being such a sap. *Get a grip, Bové, nature's rough.* But the hormones of pregnancy are potent mojo, capable of exploding even the smallest concerns out of proportion, and I had a hard time keeping it together there for a minute or two. Maybe I wasn't ready to be a parent. How was I supposed to explain to a child the ecological beauty of a world where an otter had to rip a living chick from its egg in order to eat? Of course, that was just the way of things; the system worked perfectly. But it wasn't kind. It wasn't easy. And for all the cruelty that occurred in nature, there were worse things that dwelled in the hearts of human beings. I wasn't prepared to teach a kid these realities. I was still trying to figure them out for myself.

I took a deep breath and shook my head to clear it. *Enough.* The cranes weren't wasting precious energy crying over life's injustices, and I sure as hell wasn't doing anybody any good by moping around out in the middle of the marsh. Whatever the future held, one thing was certain: if I had any sense, I'd get back to the truck before I screwed up and swamped my waders. The rest would work itself out.

I patted the small bulge of my belly. "Let's go, little one."

There was a growl within, and another bubble of queasiness rose toward my throat. "Yeah, I know," I groaned. "It's time to find some lunch."

32

Eating Crow
(and Other Ways to Atone for Sins in the Wilderness)

CHRIS SMITH

In late October 1970, with a new wife (of three months) and a new baby (of two months), living in a one-room efficiency apartment in Fairbanks with no visible means of support beyond the twenty-five dollars a week I was getting for playing guitar at a coffee house Friday nights and an ROTC scholarship, I thought a quick trip to the Taylor Highway to cache some caribou meat for the winter seemed like a hell of a good idea.

The fact that I was carrying twenty credit hours that kept me on campus from seven-thirty on Monday through five o'clock on Friday simply meant the trip had to be a no-nonsense, down-and-back affair. I didn't have the luxury of going on a speculative, leisurely hunt so I kept in touch with the guys down at Fish and Game and planned to make a weekend dash to Chicken Ridge when the peak of the migration was on. Of course, there were only about 5,000 other people in Fairbanks with the same thought, 4,827 of whom were between the ages of eighteen and twenty-three, raised in rural southern families with a long tradition of killing things and fresh out of basic training, where their inherent skills with firearms had been honed to excess. Or was that prone to excess? Whatever.

When the right weekend came along, about the end of October, a classmate and I tossed our hunting gear into the back of my '65 Ford pickup and headed to Tok. I should mention that my classmate, Tim, was also a wildlife management major and was taking Gerry Schwartz's parasitology class. As a requirement for the class, each student had to necropsy five different animals to identify all the ecto- and endoparasites they could find. Tim figured he could not only get some meat but fulfill 20 percent of his coursework to boot, all for the price of a tank of gas and the loss of a few nights' sleep.

We left Fairbanks about six o'clock Friday night. The drive to Tok was relatively uneventful, if you overlook sideswiping a buffalo about six miles east of Delta Junction just after nine o'clock. It wasn't hard to overlook the dent he left in the passenger side of the truck...you couldn't really tell it from the dozens of others that were already there. By just long enough before daylight to catch about fifteen minutes of sleep on Saturday morning, we'd made it to about 105 mile on the Taylor Highway and found a spot about three hundred yards long to squeeze into the firing line between a group of GIs from Wainwright and three or four Indians from Tetlin who had somehow coaxed the oldest Chevy I'd ever seen to the top of the Tanana Hills.

As the sun rose over the Yukon to the east, we glassed the ridges to the west, eagerly searching for the waves of caribou the Fish and Game guys at the check station at the junction had assured us we'd find. (Given the accuracy of that prediction, you'd think I'd have had better judgment than to take a job with the outfit a few years later and stick with 'em for a couple of decades.)

Suffice it to say that after about two hours glassing, followed by three hours hiking around to get a better vantage point—to no avail— we decided to head to lower country to try our luck and get out of the wind. We drove back down to around 70 mile, where it was calm but about minus 25 and dropping as fast as the sun at three-thirty. I pulled over at a wide spot in the road and got out to heat up some water for tea or instant soup to cut the cold.

I had the Svea stove set out on the tailgate with a one-pound coffee can of water set on top to boil when Tim spotted a lone raven flying slowly down the road. Given the lack of other opportunities to fulfill his collegiate obligations on the trip so far, he snatched his .243 from the cab of the truck and peeled that raven out of the sky like it was something you'd do any day of the week. The truth is, neither he, nor anyone else I know, could do that twice in a lifetime. Nevertheless, with that one lucky shot he was 20 percent closer to an A in a required class, so he started into necropsying the raven with the intensity of a brain surgeon about to open a skull.

Now, for the uninitiated, to do a thorough necropsy for parasites, you have to start by carefully picking through the feathers or fur to look for mites, fleas, nits, or similar hitchhiking life forms. Tim was about midway through the process, with the bird laid out spread-eagle, so to speak, on the tailgate next to the stove when a newer-model pickup, driven by an older couple, passed by.

As was customary in the days before making eye contact with a stranger on the highway could get you shot, they glanced our way and nodded as they passed. Tim and I each gave a weak wave back but probably weren't very expressive, since any extra movement displaced what minimal warm air our bodies had created under our cheap coats.

The truck drove on for about two hundred yards and then came to a gradual stop. After it idled for a minute or so, the backup lights came on and the old couple inched their way back down the road until they stopped beside us.

There was a brief conference between them, then timidly, almost apologetically, and certainly sympathetically, the old gal rolled down her window and said, "Boys, are you hungry?"

It took a few seconds for our sleep-deprived, cold-numbed minds to figure out that not everyone who saw a couple of ragged, half-frozen kids standing behind a beat-up truck in the middle of nowhere with a pot of boiling water and a half-plucked raven would recognize science in the making. Whether it was indignation that their generous offer was ignominiously refused, or terror in response to the hysterical fit of laughter that overcame Tim and me once the meaning of her question sank in, the old lady rolled up her window and she and her husband roared away in a cloud of ice fog.

33

Chasing the Tail of a Frog

ELIZABETH DAYTON

It was late summer, the season of fire, and the hillside across the valley from me was dancing with flames. The forest had been cleared, the good wood hauled off, and now what was left was tinder for a terrible bonfire that rose to the heavens in thick, black plumes of smoke. Where once large Douglas firs had whispered verdant secrets in the wind, now there was only raw earth and amber coals to fuel the fire. All that carbon locked up in the dead wood was escaping into the air as carbon dioxide. After working in the cool, clear streams of northern California beneath the shady trees all summer, I felt almost sacrilegious watching the vestiges burn. Yet I could not look away. Hypnotizing flames licked the sky, teasing the horizon. I could almost feel the heat from where I stood.

I turned away from the flames before it got dark, scrambling down the steep embankment to the small tributary below. I could smell the acrid scent of wood burning, but otherwise down here I was oblivious to the destruction around me. The headwater creek flowed from the mountains beneath a cover of thick-crowned trees. I'd spent the day in that stream, searching for larvae of the Tailed Frog, *Ascaphus truei*. Marbled gray like the rocks and hard to see if the light isn't hitting them just right, these tadpoles attach themselves to the cobbled stream bottom, hanging on by suction-cup discs on their mouths. Whenever I found one, I'd dip my hands into the numbing water and pull it off the cobble to break the suction and flush the tadpole into the net that I held downstream. These discs allow the tadpoles to maintain their position in the swift current while they graze on the algae that covers the rocks. If it weren't for this adaptation, they would be flushed downstream to the hungry jaws of a salmon. You almost never find these frogs in streams that are inhabited by fish.

I bounded down the creek bed, my spiked rubber boots gripping the mossy logs and slick rocks that jutted from the water. I could get going pretty fast with these boots, even with a backpack full of gear and a bucket in my arms. The three-hundred-meter length of string that I had used to measure my distance up the creek hung limp among the bracken ferns and tiger lilies. Gathering it at my side, I periodically stopped to untangle it from the thick vegetation before continuing along the watercourse. The steepest part of the stretch was a cascading waterfall lined with slick bedrock. On the way upstream, by jamming my fist into a crack, I had managed to find enough footholds to raise my body up over the vertical edge to the side of the main flow, but coming back down was a different matter. The shiny, wet surface reflected the sky, revealing no solid hand or footholds. Covered with a tangle of poison oak vines, the embankment beside the creek rose steeply on either side. There was no way to climb out of the streambed unless I wanted to be covered in oozing blisters that kept me up at night, scratching until my sheets were spotted with blood. I decided to brave the waterfall.

With my cheek pressed against the cool bedrock, I inched down the vertical surface backward. Unable to lean far from the rock wall for fear I would knock off my center of gravity, I had to rely on my sense of touch to lower my feet slowly, one by one, down the waterfall. The sound of the water traveling over the sheer edge was a white roar in my ears. The spray misted my body, and each hair stood up on my arms, despite the great heat that was generated from my gripped muscles. Cautiously, I continued my descent until I came to a point where I could feel no more footholds to climb down and I couldn't reach the crack on my left to pull myself back up. I paused for a minute, legs shaking, eyes darting wildly around for a way out. My boots were still ten feet from the step-pool below that swirled with white water. It was too far to confidently jump, but my toes were slipping from the meager indentation they had found. My right leg slid out from beneath me, and the motion knocked the left one off balance so I hung trembling by a tiny lip in the rock. All at once my arms gave out and I fell, plunging down into the pool, which was deep enough to soak me with frigid water but not quite deep enough to cushion my fall. I landed on my left ankle and heard a loud popping sound in the joint as my body came down on it.

Intense pain seared through my lower leg, and I thought for sure it was broken. The cold water took my breath away so I crab-walked out of the whirling pool, then held my ankle as I grimaced in pain. I was two hundred meters from the truck, and I was going to have to crawl out through the poison oak vines and stinging nettles. I carefully took off my water-filled boot, removed my sopping wool sock, and examined the injured ankle. It was swollen and a little purple. I dipped it back in the icy water for as long as I could stand to stop the swelling. It was already almost dark in the streambed; rosy streaks of cloud laced the sky above the trees. I reached in my pack to find my headlamp and caught movement out of the corner of my eye. There on the bank, beneath a spray of salmonberry, was a Tailed Frog.

He was thin and wiry, shining coppery like a new penny in the glow of my headlamp. His hind feet were webbed, and on his forelimbs I could see thick, black nuptial pads that almost looked like the blood blister I had on my thumb. These pads are used to grasp a female frog so she cannot slip away during mating. His face was smooth where most frogs have a translucent tympanum. Tailed Frogs have no need for eardrums in these noisy, rushing creeks—nor do they need vocal sacs. Who would hear them if they called? I turned him over and examined the thin "tail" between his legs, this species' namesake. It is not actually a tail but a unique organ used for fertilization; it engorges with blood like a penis and lodges into the female's cloaca, where it deposits the sperm. The female, which can store the sperm for up to two years, attaches a pearl-like strand of fertilized eggs to a rock in a sheltered part of the stream.

The Tailed Frog is one of the few frogs to have internal fertilization and the only frog to have an intromittent organ to deposit the sperm. I once found a male and female Tailed Frog in amplexus. I spotted them in the clear waters of a pool, and in my excitement I accidentally dislodged a rock that was damming the pool in place and the pair went crashing downstream. I located the couple in the next pool down—

still in amplexus! It became vividly clear to me why this species has evolved the "tail." Without internal fertilization, the glass-like beads of eggs would wash away into the watershed before the wriggling sperm cells could ever reach them.

I measured the frog, recorded some notes in my field journal, and then gingerly lifted my body in the twilit creek bed. Though it throbbed with pain, my ankle could bear my weight, so I knew it probably wasn't broken, just sprained. I wrapped it tightly in a long-sleeved shirt I had in my pack and slowly limped my way to the road. When I finally reached the truck, the sky was pitch black with no moon and I could barely see Scorpius, lowering on the southern horizon. Still a couple hours from home, I was thankful my work truck was automatic so I could relax my sprained left foot while I drove down the sinuous logging roads with my right.

I woke up late the next morning to the sound of NPR on my clock radio, something about wildfires in New Mexico that I'd incorporated into my foggy dream. I bolted out of bed, wincing with pain as I put my full weight on both my feet. My wounded ankle was a strange greenish-purple color. I wrapped it in an Ace bandage and pulled on my stiff but dry Levi's, jog bra, and a clean flannel shirt. If I didn't get an early start I'd never finish the creek I needed to sample today. And I'd never been to this one before, so I'd be using aerial photos and a compass to find my way. The landscape changes with time as men shave it and shape it, roads collapse into creek beds, and mountains fold upon themselves in great mud slides; it's more like putting pieces of a puzzle together than following a map.

I pulled up in front of the local food co-op in the company truck to get coffee and a scone for the long drive up north and some lunch items I could throw into a bag. During these twelve- or fifteen-hour days I barely had time to eat, shower, and sleep, let alone shop for groceries. Outside stood a tall, skinny hippie kid, his spine curved over like a cobra waiting to strike. He looked at me from beneath an army-issue cap with steely eyes that riveted back and forth between my tiny frame and the timber company logo on the side of the Ford I was driving.

"Good morning," I said, a little intimidated and trying to melt the glacier that was forming between our eyes. He chuckled a little, spine reeling backward, hands shoved into his pockets as though holding on to some inner strength. He walked up to me and I thought he was going to shake my hand, but then he coughed a deep smoky cough, mustered up some infectious slime, and spat the yellowish mucous near my feet. I stood there for a moment in awe; what do you say when someone spits at you? My mind wandered back to kindergarten to think of the proper etiquette in such a situation. Sticks and stones may break my bones, but what about mucous and saliva?

He squinted his eyes, laughed a guttural laugh, and hissed, "Blood money! You're living off blood money—the blood of trees!"

"Excuse me, you don't even know what I do," was all I could manage to say—here I was being polite to a guy who had just spit at me.

"Oh, I've seen enough of you around. You probably go out and tag the trees, telling 'em which ones to murder. Only thing worse than a hit man is the one giving the orders."

"Actually, I'm a biologist—a herpetologist to be exact. I work for the timber industry, studying frogs and salamanders, monitoring the populations to make sure they aren't being negatively impacted by timber harvest."

"It's all just a scam. Your paycheck still comes from the blood of trees." He laughed as if he had just said the most intelligent thing in the world.

"Well, at least there's someone out there monitoring what's going on. Ten years ago the timber companies rarely hired biologists to keep track of environmental impact. Some companies still don't. The way I look at it, I'm doing something worthy by monitoring the Tailed Frog. My ultimate goal is the preservation of that species. Can I ask, sir, what do you do?"

He squinted at me with those steely eyes and in a hateful voice said, "You're nothing but a biostitute!" Again he shouted, "Biostitute!" so that the early morning working crowd, as well as the all-night loiterers still in the parking lot, all turned to look at my small body quivering in one tennis shoe and one stocking foot.

"Well, if you'll excuse me I have to get off for work—unlike you, I suppose," I said as I limped into the co-op.

The word *biostitute* stung, and I pictured myself limping on stiletto hills instead of an injured ankle, wearing a vest filled with spotted-owl down, a stole of fisher fur around my neck, epaulettes of Marbled

Murrelet feathers displayed on my shoulders. A prostitute exploits her body for money; a biostitute exploits life for money. I'd heard this slang word before, used to refer to a biologist who fudges the data or turns a blind eye to environmental destruction when handed a large sum of money. No, I thought to myself, that is not what I am.

My hands shook as I handed money to the counter attendant. I could feel my cheeks flushed, burning, and I asked myself, "Why didn't I tell the guy off?" As I waited for my coffee, I grabbed a handful of napkins and felt a pang of nausea as I wiped the side of my tennis shoe to be sure none of the saliva had splattered me. The thought of this guy's bodily fluids on *my* body made me sick. I decided to go back out there and give him a piece of my mind. I felt like a lion in a house cat's body, a tempest waiting to explode. When I got out to the parking lot, he was gone.

On the way up to my field site I drove through the national park, one of the last stands of old-growth redwoods in the world. The treetops were jagged and uneven with crowns of various ages. Ragged-looking snags poked through the tree line, and just as I gazed out my window I saw a Bald Eagle soaring overhead. Tiger lilies bloomed everywhere, vibrant orange against the cool green of redwood sorrel that lined the forest floor. The big trees just stir up something inside you deep and almost religious. Perhaps it's partly that they are so big: you feel as if you're standing in a cathedral looking right up at God himself. But I think it is more than that. When you stand next to those 350-foot trees that are wider across than three of you lying head to toe are long, it makes you feel very mortal. It is humbling to think that a tree has been standing there since before you were alive, before your grandparents or great-grandparents were alive, and in some cases even before the time of Christ. If these ancient beings could talk, what wisdom they would impart.

After the park I drove through miles and miles of Forest Service and timber company land. Some areas were lush with emerald forests, a sea of trees. But in other areas the stubble of clearcut scorched the earth like razor burn. A logging truck zoomed by me on the interstate, leaving a trail of shaggy redwood bark that flew off the felled trees as the rig sped up the highway. I marveled at the size of the logs poking off the back—you hardly ever saw them that big anymore. The diameter of each of the boles was so large that only three could fit on the truck. I noticed the bumper sticker the driver had plastered to his rig: "Save a tree—wipe your ass with a spotted owl." My knuckles whitened on the steering wheel, and I shook my head back and forth.

"Why does the whole issue have to be so polarized?" I said to myself out loud.

Most of the people I'd met in this place were either "tree killers" or "tree huggers," with hardly anyone in between. It burned me up that the timber industry and the environmentalists couldn't work together consistently to both generate board feet for timber *and* maximize biodiversity. On the one side are the extreme environmentalists who pound spikes in trees, potentially killing the logger who is cutting the tree. On the other are the staunch anti-environmentalists who do things like "ring" old-growth trees in the national park lands—cutting around the tree through the cambium for no other purpose than to kill a thousand-year-old tree. I did not understand why these people refuse to work together when they both want, and need, the same thing: trees. All this hatred just didn't make sense to me.

Once I'd pulled off 101 onto the dirt logging road, I hopped from the truck on my good foot, unlocked the gate, drove through it, and locked it up again behind me. I had to drive carefully at this point, comparing the bends in the road with those on my aerial photograph so I'd know where to pull off and hike down to the site. I reached a tiny spur road that tailed off to the right, and I was pretty certain it was the road I wanted, so I jerked the truck into four-wheel-drive and bumped down the boulder-strewn path. I could tell it hadn't been used in a while; grass had grown in tall and obliterated all previous tracks. Suddenly, the road came to a halt, and I slammed on my brakes right before I drove off a fifteen-foot drop into a gully below. Adrenaline pumped through my racing heart as I got out of the truck to investigate. The road had simply washed away; it must have happened during the big storms we'd had in the spring. I got out the sequential aerial photographs of this area and studied them to figure out where I was going. Luckily, I'd brought along a small stereoscope; by holding two overlapping photos together beneath the device and focusing the two images as one, I could view the landscape in topographical relief. I saw that I would have to hike down this road about a quarter mile after the washout (which was not shown in the photograph, of course), then cut through a steep clearcut area to the lush riparian buffer zone. I marked on the photos with a red grease pencil to show myself where I'd have to descend and carefully placed them in my backpack.

I forced my swollen foot into my rubber boot, wincing with pain, then faltered around the washout and down the shady road through the long grass until I came to the clearcut. The burned stumps blackened my jeans with charcoal strokes as I shimmied down the steep slope. Heat pounded my shoulders, and sweat began to dampen my shirt. The charred wood had the ashy gray look of old bones, a graveyard for trees. Someone had come in and planted seedlings every yard or so, each one carefully wrapped in plastic mesh to protect it from deer. They were about a foot tall, hope for the future. The area must have also been sprayed with a herbicide, though, as there were no alders, blackberries, or even wildflowers growing on this ashen plot of land. I envisioned the monoculture of Douglas fir trees that would regenerate, a forest deprived of flowering dogwood in the spring and the habitat complexity that comes from a mixed stand; this would be a forest vulnerable to damage from insects and disease.

The forested zone that traversed the creek was a cool refuge from the hot, barren clearcut. As I entered the green light that poured from the canopy, my damp shirt felt cool on my back. I could smell the pungent scent of California laurel. The forest was a diverse mixture of Sitka spruce, Douglas fir, tan oak, and Western hemlock. A dead fir tree stood in a small clearing, the bottom hollowed out by a bear, the top a jagged fork littered with woodpecker dens, perhaps long vacated and since inhabited by owls, bats, songbirds, squirrels, martens, or other species. The trilling song of a Warbling Vireo tickled the air, and I took a deep breath of the stillness. Carefully, I stepped over an enormous fallen log, and something shiny caught my eye. It was a Clouded Salamander, curled up in the bark. I picked up some of the bark, and it broke apart and fell to the earth, nutrients returning to the soil. I peered closely at the rotting tree, a maze of termite tunnels, fungus hyphae, and small hollows filled with rodent droppings. Licorice ferns grew from the mossy bark, and numerous tree seedlings had colonized the humus that crumbled from the dead tree. A fallen log is the coral reef of a forest ecosystem.

As I neared the creek, I could hear the rush of water tumbling over rocks and cobble. A raven resting on the mossy limb of a cedar vocalized a garbled sound as I approached the water's edge. From here I had to hike a few hundred meters upstream to the place where the main channel is fed by two smaller tributaries—the headwater streams that drain from rock-filtered seeps deep inside the mountain. About ten meters up from the confluence, I found what I was looking for. The tadpoles were attached to the rocks by their powerful suction-cup mouths, little white-tipped tails whipping in the swift current. Their bodies were cryptic, ranging in color from black and gray to brown and tan, speckled to perfectly camouflage with the surrounding rock surfaces. It had taken me a few weeks to get their search image patterned in my mind so I would not miss them as I scanned the creek.

These animals were just outside my stream transect so I left them alone, securing the downstream boundary of my sampling area with a net held in place by rocks, sticks, and roots. To search the creek, I used a "viewing bucket," a plastic bucket with the bottom cut out and a glass "window" secured in its place with clear caulking. With this tool I could see beyond the silvery reflection of the sky chopped up in the riffles to the ruddy creek bottom filled with gravel and cobble, Pacific Giant Salamander larvae, and the occasional Tailed Frog tadpole. With a dip-net in one hand and my viewing bucket in the other, I hunched over the creek all day, following a strict sampling protocol that would give us an estimate of the Tailed Frog population in this creek. By comparing the numbers of animals I found in this creek with a "control" creek (a creek as identical as possible in every variable—slope, aspect, geology, canopy cover, vegetation, etc.—except for the fact that no trees had been logged in its drainage), we could get a reasonable indication of how timber harvest was impacting the population of Tailed Frogs.

The larvae I found were second-year larvae, last year's brood, getting ready to turn into little froglets. Their back legs were lengthening, and the forelimbs were beginning to emerge from the slits beneath the operculum, where they formed in safety like a baby in the womb, not emerging until they were functional for locomotion. The mouthparts still had the suction cups on them, but I could see where they were beginning to change—becoming less rounded and "fishy" and distinctly more "froggy." Soon the scraping teeth would disappear, and in their place the animal would have a finely etched jaw more suitable for grasping insects than eating algae. Along with this change the intestines would shorten. I could still see the coppery coils visible through the skin of the tadpoles—long, hollow ropes twisting and meandering, providing enough surface area for the absorption of nutrients from the green algae that covered the surface of the rocks.

Hunkering down in the streambed surrounded by sword ferns and closed in by a canopy of enormous conifers, I imagined I was back in time 150 million years ago when pterodons flew through the air. The Tailed Frog is a relict amphibian species, an ancient survivor from a previous time. It is thought to have been widespread during the Jurassic period, when dinosaurs roamed the earth, but changing climate and harvesting of the big trees has restricted it to this cool, forested fog belt of the Pacific Northwest. This emerald green sanctuary has acted as a refuge for these descendants of an earlier era. If the trees were cut down and the temperature in the creek bed was raised to that of the land in the clearcut above me, the Tailed Frog would most likely be lost forever. It dawned on me that I was working on a living fossil, a species that was hanging on by the last of its evolutionary threads.

Many people might ask, why should I care about the survival of the Tailed Frog? Will it really affect my life if it disappears forever? The truth is, we don't know what would happen if the Tailed Frog were gone. Perhaps it serves an integral function in the riparian ecosystem. Perhaps without the Tailed Frog tadpole the algae would grow out of control on the rocks; without the adults the insect population would certainly amplify, sending reverberations up the food web. Perhaps the Pacific Giant Salamander, whose larvae feed heavily on Tailed Frog tadpoles, would suffer.

But let's just say that it wouldn't change a thing about the ecosystem, that Tailed Frogs already exist in too small a number for their disappearance to have an effect on anyone. I would still have to argue that the survival of the Tailed Frog *does* matter, that we all should care. This species has an intrinsic right to remain on this planet because of the fact that it is here. The Tailed Frog is the unique product of millions of years of mutation, recombination, and random chance, honed by natural selection to create an amazing creature that is perfectly adapted for the stream environment in which it has evolved. This species is connected to us, not only by the genetic language of nucleotides that we share but also because of the fact that we both depend on the same planet for our lives. We *all* need the trees—loggers, environmentalists, scientists, and Tailed Frogs alike. We are all connected. What gives us the right to choose one species' comfort over another's preservation?

The sun had almost set as I emerged from the creek far upstream in the watershed. Ambling slowly through the tan oak brush, I heard the squawking of Steller's Jays above me. From the ruckus they were making, I knew there must be an owl nearby. I tilted my head back, took a deep breath, and let out a loud, hollow call: "*WHOO-whoo-whoooo-WHOO!*" She swooped down to a low limb on a Douglas fir beside me, a spotted owl with inquisitive yellow eyes. We stared at each other for quite some time, and finally she looked away, searching the dry carpet of tan oak leaves and fir needles for a pack rat or vole. Another owl called from somewhere deep within the forest and her head perked up, turning toward the direction of the sound like a sonar receiver. I felt the cool air from her wing beats lift my hair as she flew silently away.

I came out from the hushed forest to a prairie of thick, green grasses, dappled with lupines in the muted light of the sunset. I found a low-limbed prairie-intrusion tree, a wolf tree, standing alone at the edge of this meadow of wildflowers, and sat beside it to watch night descend on the quiet clearing. I gently removed my boots and wet socks, unwrapped my ankle, and elevated it on one of the tree's mossy limbs. Mountains surrounded the prairie, displaying swatches of clearcut areas like a patchwork quilt. I could see the fresh cuts and those that had been burned to a black, ashy crisp. Some areas were replanted, appearing green even in this muted light. Still other patches were thick and dark, so lush I could almost smell the spruce and fir from where I sat. The human hand has done much to change the landscape of this earth. Since we are not going away any time soon, and our population is not getting any smaller, the trick will be to find a balance in the near future—a balance between what we use and what we replenish, a balance between human inhabitation and the preservation of other species.

I sat still for a very long time that night, my ankle throbbing, my lower back easing into its usual dull ache from bending over the stream all day. The moon had not yet risen, and the sky was a pale indigo blue, sparkling silently with the first few stars. In the soft light the prairie looked vast and smooth, a continuous sheet of velvet that rolled from grassland to mountains to sky with no spaces in between. In this light I couldn't see where the cuts began and ended; it was all one unbroken landscape that was peaceful and still, with no human for miles but myself, and I blended right in. I heard coyotes howl in the distance, playful vocalizations that lifted the hairs on the back of my neck and made a tingly sensation travel across my scalp. I looked up just as the waning gibbous moon rose over the horizon, silhouetting a jagged tree line of ancient conifers and broken-topped snags. Stretching my body out as if to feel the warmth from the moon, I smiled.

34

Dances with Coyotes

J O H N   S H I V I K

"No shit, there I was."

These kinds of stories always start that way, and that's exactly what I was thinking. Same old hyperbolic same-old, but there I was with a large male coyote—at least 102 pounds—with his canines sunk into my thigh. The glamour of being a wildlife biologist.

My next thought was, "What the hell should I do now?"

It all started out innocently enough. Master's project, U.C. Berkeley. The Sagehen Research Station up in the Sierra, only a thirteen-mile ski from Truckee. Had to catch coyotes and put them on line for monitoring. Had to get a degree. Had to change the world. But how was I going to do that when a coyote's fangs were dangerously close, snapping at the only thing in the world more dear to me than the world? I had to think.

No kidding, there I was—but we covered that.

I had been trying to capture coyotes for about a year by then. I think I had put sixteen or so collars on, but I was halfway through my degree program and about half the collared coyotes had either dispersed or been killed. My career as a wildlife biologist was in danger of being found dead on the roadside. I had to find a way to make it work.

The idea was really simple. Put a few radio collars on coyotes and watch what they do up in the high mountains. Other people claimed these animals weren't territorial and that they migrated like swallows down from the mountains in order to find greener pastures below, especially those filled with sheep.

Wrong.

In spite of being an uppity, self-righteous academic, I knew that if I wanted to present a solid argument regarding coyote behavior and to change the *spiritus mundi*, I needed data, but at that moment I was I wasn't feeling particularly academic. One's noble ideas fall quickly by the wayside when sharp pointy teeth are snapping inches from one's genitalia. No, at that moment, my drive was primal—*mutually assured destruction*. If my ability to reproduce was taken from me, well, as caveman as it sounds, I sure as hell was going to take his from him.

So there I was with my whole graduate-school life flashing like pearly sabers before my eyes. I thought that my study was important. I thought that I was somebody and that I was going somewhere. I figured that if I could just get a few more collars on a few more coyotes, then I'd be able to save the U.S. government and prove to the Evil Empire, then known as Animal Damage Control, that a better understanding of predator ecology was the only way to develop sane methods to counteract the dangerous proclivities of predators. I knew I could reason it out biologically.

Screw that. Right then, the important issue was that I had captured a big male coyote that was much more obstreperous than they usually are (would you believe I actually used the word *obstreperous* in my field notes that day?). Most coyotes are really pretty easy to handle. When they are cornered, really cornered, they simply flop on their sides and assume a subordinate position. Tail curled up, not making eye contact; they use social cues to indicate surrender. Maybe it says something about trans-species communication: the human ability to domesticate dogs and their ability to compel us to love them. With most coyotes I put a collar on, I feel like we understand each other.

There was the coyote I had captured out at Mono Lake, for example. Because of the high desert summer and the difficulty getting out to the island where I had set the trap (fully confident that no coyote would ever show up there), the poor animal was captured and held until later in the morning than I'd have liked her to be. She submitted as she was supposed to, and I muzzled her and did the usual processing like measuring from nose to tail, measuring hind foot length, and all sorts of things that seem incredibly inane to me now. But I was also trained to monitor basic life signs.

I frequently and unceremoniously inserted a thermometer into her rectum (sensitively lubricated with KY, because I am not a barbarian). Her stress became evident as I watched the mercury rise: 101, normal. 102. 103. Holy shit. 104. 105. I dropped everything. She was getting too hot in the morning desert sun. I wasn't done measuring and processing, but I had to stop. I couldn't kill my study animals. The baby is more important than the bath water, I thought, and suddenly I realized, "Bath water! That's it! Bath water!" I scooped her up in my arms and carried her to the cool waters of the lake. We walked through the desert sun, and I waded in like John the Baptist and dipped her into the cooling waters, massaging the liquid over her ears, around her neck, and into her scalp. And then it was like magic. She looked into my eyes in a way that was close to First Love. We connected. She felt better immediately. She trusted me. She smiled.

Thoroughly doused, I carried her back onto the island and found a bit of shade near a large rock. I removed the binding Velcro-like veterinary wrap that I had used to secure her legs. I carefully unwrapped her muzzle, thinking, "She's got to be pissed," but she didn't try to run away. She didn't try to bite me either. She just burrowed into the shade at the base of the rock and nestled in the cool sand, watching me. I gathered up my things and retreated back to the mainland. She survived, did fine actually, rejoining her pack shortly thereafter, and I learned a valuable lesson about animal care and where data come from.

Unfortunately, that experience did not in any way prepare me for the situation that I found myself in with the big male—holding the tail of a coyote that was more crazy and uncontrollable than a two-year-old on Christmas Eve. What had happened was this: when I first pinned him, he fought a little, but I was able to grab and wrap his muzzle with vet-wrap. I secured his legs, tying the front two together and the back two together. He rested, hobbled for a moment. Then I began to process him. I don't know, maybe it was the thermometer thing, but when I stood up to get the socket driver for the radio collar, he used the moment to make an all-out effort to take control of the situation, to prove he was the alpha here, not me. Lying on his side, he brushed his muzzle across the ground and swept the sticky wrap off as if he was throwing away a used Kleenex. At that moment the world staggered and slowed. I could hear my voice wailing an artificially baritone "*Noooooooooooooo*" as if someone had put the videotape of my life into frame-by-frame mode. I reached down incredibly quickly, but time was moving so slowly that it seemed to take about forty-five minutes to get to him. The good part was that I was able to grab him before he was able to hop away. The bad part was that the end I came up with was not the business end. Worse yet, he was not interested in running away. When I reached down and grabbed his tail, he hopped up like a bipedal berserker and snapped his fangs into my thigh.

And there we were, me holding his tail and he hopping up repeatedly trying to sink his teeth into my flesh again. We danced, the alpha-male stud coyote and the researcher who had set out to find a better living through biology. Man, he was pissed. But I was above that. I was more focused, more hungry, more poor, more motivated. Damn it, I was a graduate student.

Because I had him by the tail, and his legs were bound, he was really limited in what he could do, although he had managed to get his muzzle off and was able to leap up and make a snack of my chicken legs. As you could probably guess, I was in a position of serious scientific conflict and ambivalence, the ramifications of which had not been seen since the pope imprisoned Galileo. No! This was more serious than that . . . since Giordano Bruno was burned at the stake! This coyote, this wound, this dance was bigger than me. This was for Science.

If I acquiesced to the threat of bodily harm, my thigh—surely my genitalia—would remain intact, but I would have to give up that most important goal of the graduate student. More important than beer, more life-giving than ramen: *sample size*. I needed all the coyotes I could get in order to validate my study. The irony stung like twin bodkins. Size does matter.

We performed coyote ballet that day. I held the base of his tail and suspended him above the ground just high enough to keep his lunges toward my groin ineffectual, but not so high that I was holding all his weight on his spine and perhaps hurting him.

So we went around in a circle three times, as any dog would do, and then simultaneously realized the futility of our struggle. We achieved détente when, in a mad attempt to stave teeth away from my important bits, I tried to pin his muzzle to the ground with my foot. This was a satisfactory concession to the coyote because he used the opportunity to authoritatively latch his canines into the sole of my boot.

"Ahhhh," I said. "Progress."

There we were. The coyote used my Vibram like a rawhide, and I gently suspended his rear end above his body, holding enough leverage just in case. Eventually, I was able to bring my other foot forward and pin his neck down. He was generous enough to let me do this because he still had the first foot in his mouth. I was happy because, due to the solid construction of my boots, my vermiform appendages were undamaged. With amazing balance, I contorted like a champion yogi to reach gently between his ears, over his forehead, and around his muzzle. Finally, I had his mouth. Things were getting better. Kneeling over him, I extricated my foot and swore off vet-wrap forever. This was a job for electrical tape.

I put a collar on him and took a few more inane measurements. Then I let him go. He looked back a few times. He peed on a bush in front of me. At last he loped away.

With my task safely finished, I dropped my pants and used gauze from the vet kit to treat my wounds. I kept direct pressure, using electrical tape. I limped my way to the hospital a few hours later and tried to explain to the doctor why one living coyote was more important than the post-exposure rabies vaccinations I was about to receive, but he didn't understand.

A couple of weeks later the big male ran off and died on the roadside anyway. Ultimately winning, he left me data-less. We understood each other, all right.

So that's the story. I resigned myself to the fact that I wouldn't gather the necessary sample size to change the world. I realized that there was no solution and no Evil Empire, just coyotes and people doing their natural dances and trying to figure out how to survive with, and despite, each other. Even if I ended up a failure as a biologist, the dance would go on.

Was it all worth it? The bite. The incredibly painful wound irrigation with rabies antibodies. The repeated indignity of 20 cc of inoculate into each butt cheek—not to mention how difficult it is to sit and drive into the field with that much extra fluid on board. I don't know, but looking back now, after all these years of research efforts won and lost, I just pat the faded scar on my thigh and remind myself that at least I still have my testicles.

35

Snapper's Kiss

JEFF BEANE

You could call it a case of mistaken identity gone terribly awry. Normally, I wouldn't have been in such a hurry, but it was one of those "friendly competition" scenarios. Jim Warner had set up a video camera atop the hill overlooking the wet pasture harboring a population of Bog Turtles that we were studying. Both rare and secretive, Bog Turtles can be hard to find, and most folks hunt them using a combination of visual search and probing the muck with a broom handle until they hit one. But techniques differ. Some people are primarily visual searchers, some nearly exclusively probers. Some work the habitat randomly, some selectively. Some use one stick, some two, some none. By filming everyone together and studying the video, Jim hoped to analyze and learn from the different techniques. Everyone wanted to be first to find a turtle for the camera, and we were competing against some of the best Bog Turtle hunters in the Southeast.

En route to the "good" part of the wetland, I passed through an outlying section of mostly hard-packed mud with water-filled ruts and pools. I jabbed the water pockets quickly as I passed—you never know where a Bog Turtle might be. When my stick made contact with a hard, hollow, nonstationary object, I knew it was either a Bog Turtle or a small Snapping Turtle. We often probed snappers while hunting Bog Turtles. Usually, it was easy to tell the difference. Adult snappers, being much larger, sound and feel different. But small ones would sometimes fool us. If this was a snapper, it was a little one.

Or so I thought. It was a snapper, all right, but I had no way of knowing that the great majority of the animal was nestled in a pocket beneath a hard bank of mud. What I had probed was only its head. By the time said head had been thoroughly agitated by Orange Crush (we sometimes named our Bog Turtle sticks in those days), it was more than ready for the next thing that came down to meet it. Assessing the situation far more quickly than my brain, my hand jerked back, but not before the extreme tip of the turtle's powerful, horny-beaked jaws shredded my right index fingertip. I couldn't help but imagine what my turtle-hunting technique would look like on Jim's playback as I hurried back to the Bogmobile for the biggest Band-Aid I could find.

My most-used finger bandaged, I was slightly handicapped, but that didn't bother me nearly as much as something one of my companions said: "Man, I would have jerked that [perhaps mercifully, I fail to recall the precise term he used for the turtle] out of there and slammed him against a tree!"

Now right away I saw several things wrong with this line of thought. First of all, we were there to begin with because of our alleged love of turtles. Second, snappers are not humans. Altogether lacking morals, they can no more plot revenge than can a rock. Their behavior represents genetic programming, not meanness. Threatened, a Box Turtle withdraws into its shell. A deer runs. A skunk sprays. A possum plays dead. And a Snapping Turtle snaps—it's what they do. Third, I, not the turtle, had initiated the attack, if inadvertently. And finally, there weren't really any trees in the site except for small, shrubby tag alders.

So I didn't go back to look for the turtle. I had felt enough to know that it was an adult—perhaps ten pounds, possibly larger. Large enough to have injured my hand much more seriously. I also knew that I had been kissed by a creature that even turtle lovers find difficult to appreciate, one that will never receive the respect it deserves.

Except maybe from me. The experience only increased my already profound respect for an ancient, resilient, adaptable, primitively beautiful animal with a dead-serious attitude, the will to meet life's challenges head-on, and the wherewithal to fight back against the only species representing any real threat to its continued survival. I still bear a scar and a little numb spot on that finger, but no grudge against that snapper or any other.

And about an hour later, just after Jim shut down the video camera, I not only found a Bog Turtle (with my left hand) but was the only one of the group to do so.

36

Whistling Dixie in the Land of the Midnight Sun

BENTON TAYLOR

I was stationed in Kodiak, the archetypal Alaskan town, whose lifeblood is blue and salty. As on most islands, everything there revolves around the sea. The shore was lined with seafood-processing plants, marine-parts stores, fueling docks, loading cranes, and the occasional bar in which the fishermen drank Alaskan Amber and Yukon Jack. The cool arctic air always seemed to carry the fragrance of slightly decayed fish and saltwater.

I was a month into my first contract with the National Marine Fisheries Service, collecting data aboard commercial fishing vessels as a part of the NMFS "in season" management program. The information I gathered had a direct impact on how long the fishing season would last, which translated into how much money the fishermen would take home that year. So you can imagine there weren't a lot of reasons for a fisherman to look kindly on NMFS observers. Most of the guys, however, took it in stride, and I can tell you that I never had any serious trouble out of them. Indifference or the occasional smartass remark was the worst of it, while respect and camaraderie tended to be the norm.

I stood on the deck of the *Provision* as she slipped her way back to port, and I watched the ships that were moored together in the harbor. Their masts were a forest of steel and wire undulating with the ebb and flow.

At the dock a woman named Susan, who worked for the contracting company I was employed with, was waiting to take me over to another boat. My next assignment was the *Caravelle*. I had worked on her a couple of weeks before. She was an eighty-six-footer and had originally been a shrimp boat in the Gulf of Mexico before she was converted to a trawler for the colder waters of the Gulf of Alaska. Despite her shortcomings, I had a great affection for the boat. Maybe it was because she had been built in Bayou La Batre, Alabama, and I took comfort in knowing her past resembled my own. We had both been created in Alabama, had both felt the warm waters of the Gulf splash against our bodies, and by twists of fate had both ended up in this foreign place far from Dixie.

Susan drove me over to the boat, where the three-man crew was loading up supplies for the trip. Charlie, the captain, was overseeing the proceedings. A transplanted southerner like myself, he had been born and raised somewhere just outside Dallas and was now in his late thirties. The fifteen years he had spent in Kodiak had reduced his telltale accent to the point that it was recognizable only when he said certain words. But his mannerisms and the glint in his eye said "pure cowboy."

"They say the weather's going to be up in the next couple days," Susan said to Charlie.

"Yeah, I heard about that. But don't say it too loud; I try not to let the guys find out about the weather. They are quick to start bellyaching if it gets a little rough. Besides, there ain't a damn thing I can do about it," Charlie answered through a sly grin that cradled an unlit Marlboro Red.

Fuckin' Texans, my mind sighed.

One of the annoying things about working on a boat is that space is extremely limited. Most observers working in Kodiak have to jump from boat to boat, usually staying aboard one vessel just three to five days. This doesn't allow you time to lay claim to much real estate. The fishermen, who were there first, know that you are a temporary occupant, and besides, you're just another observer. So I began the sometimes frustrating task of stowing my gear.

NMFS makes sure that as an observer, you have everything you need as well as everything you don't need to collect data, but it doesn't supply porters to help you carry it all. The list of supplies issued to me included one hanging scale, two finger scales, a flatbed scale, hundred-foot and ten-foot tape measures, a measuring board, scalpels, tweezers, a serrated knife, a life vest and a survival suit, a disposable camera, two thumb counters, a mound of various data forms, and a tackle box full of pencils, pens, rubber bands, collection vials, miniature manila envelopes, DMS (a very nasty tissue fixative), and several other trinkets whose intended uses remain a mystery to this day. All these things were packed neatly along with my raingear in two blue heavy-duty laundry baskets that doubled as the containers I'd collect fish samples in.

My bunk on the *Caravelle* was the typical rack that observers usually get. It's the one that's closest to the ship's bow, where you feel the motion of the vessel most (bunks that are midship are coveted and are usually reserved for the most senior of crew members). And if the bunks are stacked, you get the top one (it's a longer way down if you get slung out of it when a rogue wave hits). The racks are usually about three feet wide and seven feet long with the mattress slightly recessed in the frame to keep you from constantly rolling out. Whatever gear you had that you couldn't fit in your government-issued baskets, or anything you didn't want to get wet, shared the bunk with you.

I shared a room with Gordon, a short, pudgy, sixty-plus-year-old man who had spent his entire life out on the salt. He was a quiet but jovial fellow, and he had the habit of catching quick naps that never seemed to take place in his bunk. He would sleep standing in the kitchen while he waited on the coffee to finish dripping or in a doorway or sitting in the wheelhouse.

James and his younger brother John made up the rest of the crew. James, like me, was twenty-four at the time. He had spent most of his youth fishing with his father out of Kodiak. John, at eighteen, did not share his brother's enthusiasm for the occupation. This was to be his first trip, and he quickly learned what it was to be a greenhorn.

We shoved off sometime after one in the morning and began the ten-hour steam to the fishing grounds. I usually took this time to catch up on my paperwork. NMFS required us to fax in our data on a weekly basis. As I sat filling out the numerous data forms, I wondered what poor soul sat in an office and translated my chicken scratch, which was even further distorted by the rocking of the ship and smeared with whatever mystery liquids were spilled on the galley table, into useful information.

The sea was calm, and the ship rocked to the ancient motion of wind, water, and moon. I stared numbly at the pile of paper before me as the hum of the diesel engines led my mind into that quasi-conscious state that exists somewhere between deep thought and lucid dreams. The thin wisp of smoke from my neglected cigarette drifted into my face, and I closed my eyes.

She was wearing a pair of cutoffs and my old tattered UNA Lions T-shirt. Our arms were wrapped in an almost desperate embrace, and we both arched our backs slightly so we could look into each other's faces. A strand of rusty red hair had fallen into her eyes, and her tears had turned it a dark brown. She was telling me something, but I couldn't hear her. My lips began to make their way toward hers one last time.

"Time to go to work, boys! Let's go! Let's go!" Charlie yelled while he gave the traditional wake-up call by blowing the ship's foghorn.

Fuckin' Texans! I thought.

It was a beautiful September day. The sun was bright and warm, and it made you forget you were a stone's throw away from the Arctic Circle. There was a ten-foot chop and a breeze out of the southwest. Seagulls and kittiwakes hovered about, a cigarette hung from everyone's mouth, and there was no land in sight. What a wonderful goddamn place to be!

I usually had nothing to do while the crew set the net. I went through my gear and pulled out things that I'd learned were necessary: my scales, measuring board, and knife. Then I amused myself by watching the crew go about its chores.

By the disgusted look on his brother's face, I could tell that poor John had already screwed something up. He scurried around on the deck in what seemed like circles, never really getting anything accomplished. The Greenhorn Shuffle.

Gordon seemed to have another problem besides his apparent narcolepsy. Despite a lifetime spent on the ocean, Gordo appeared never to have gotten his sea legs. He was constantly falling or bumping into something. And as he picked himself up off the deck, it seemed a testament of his good fortune that he had survived out here so long, because on a ship you're just one slip away from death or disfigurement.

James was obviously the top hand despite his young age. He was knowledgeable and determined. He was always where he needed to be, and he was generally the oil for the squeaky wheel. However, he seemed to have a greater purpose on this trip than to make money. James was given the chance and the challenge to work side by side with his little brother. He had that pissed-off patience and tough-love attitude toward his younger sibling that only an older brother could.

Pollock season was over, so the crew was after flatfish and cod, known as a notoriously dirty fishery. They call it that because the fisherman drag their nets along the bottom of the ocean floor and catch anything they can, and they usually wind up catching *everything*. Jellyfish, sleeper sharks, king crabs, salmon, discarded fishing gear, starfish, and sometimes, as legend has it, remains of other fisherman who have taken the Big Drink.

After the crew had gotten the net in the water, there was usually nothing much to do for a couple of hours. So everyone would go about his business, doing the things that people do. Reading was by far the most popular pastime. The *Caravelle* was stocked with a plethora of material that included such literary giants as Steinbeck, Dickens, and Hemingway; tapered down to popular writers like Grisham, Clancy, Rice, McMurtry; and finally bottomed out with various publications by Larry Flint and a book on how to pick locks.

Charlie powered down the mains and brought the ship to an idle. Time to bring up the net and see if there was any money in it. We suited up in our bright orange raingear, slipped our XtraTuffs on our feet, and stepped out into the Alaskan afternoon.

The cacophony of the net being brought in was at full crescendo. Cables and line groaned as they wound themselves around the seven-foot-tall net reel. Hydraulics and motors whined as they pulled the net up from the depths of the Gulf. The skipper belted orders from the wheelhouse, and James yelled explanations and advice to his younger brother.

The net itself is divided into several parts. The front is mostly just heavy line and two big steel doors that are designed to drift apart with the current, keeping the mouth of the net open. The fish are collected in the codend, which looks like a traditional net and, when full of fish, takes on the appearance of a giant sausage. It is roughly thirty yards long and can hold around forty metric tons of fish. Because it is three times longer than the deck, it has to be brought on board and emptied in sections.

The crew waited in anticipation for the net to surface. Hundreds of gulls, petrels, and kittiwakes swarmed the boat, ready to snag any scraps that slipped through the net.

John hooked the gilson (a thirty-pound iron hook) onto the codend, and Gordon worked the levers of the crane until the first section was dragged onboard. James popped the hatch to the bins below deck that held the fish. He pulled out the zipper stitch and let the "slimy gold" pour out the side of the net.

I took my first sample, filling the heavy-duty laundry baskets that doubled as part of my luggage with fish. My main duty was to collect four hundred pounds of fish from each haul. I would then identify, weigh, and document each species that occurred within my sample. I set aside fifty of the most prevalent species and dissected them in order to determine their sex. Sometimes I would crack open their heads and remove their otoliths (tiny, oval ear bones) so a land-loving scientist in a lab with central heat and air could determine their age. I love being a field biologist.

We finished the haul, and the crew made another set. The sun was still shining high and bright, so I took off my raingear, turned it inside out, and left it on deck to dry out a little. This was to be the last time I'd make such a foolish mistake.

We gathered in the galley and stretched out on the black Naugahyde bench that was hemorrhaging its stuffing despite all the duct tape. James and I chain-smoked and talked about how we ended up where we were. John pined about his girl, who was waiting for him back at the docks; he had convinced himself that she was sleeping with another guy. Gordon grabbed a quick nap by the stove while he waited for his water to boil. Our world was copasetic.

Most boats have a special fax machine on board that receives weather updates from the National Weather Service several times a day. Included in the transmission are maps showing where fronts and pressure systems are located and where they are likely to track. It's meant to save lives but is used to cheat death.

"Let's shag ass, boys. It looks like we only got a couple of hours to bring in the net and steam over to one of the islands before the weather's up. With a little luck, we'll be able to anchor down in the shallows and ride it out," Charlie said. His face was tight, his jaw set. It was the first time I had seen him speak without a boyish grin.

It usually takes around fifteen minutes for the fishermen to pull the net up from its icy depths and onto the boat. So I decide to kill some time with a cigarette and a cup of coffee, which is all the time that's needed for a gale to sweep in and turn the Gulf into a chainsaw.

I went to grab my raingear and remembered that I had left it out on deck to dry. I didn't realize the severity of my mistake until I opened the door that led outside. Waves were cresting over the rails and spilling onto the deck. The boys were just standing there with "What do we do next?" expressions on their faces. Something was wrong.

I ducked behind the net reel and made my way to the opposite side of the deck where I had laid out my gear, trying to dodge the waves that yearned to touch me with their bitter cold fingers. Charlie was on the upper deck shouting out orders to the dumbfounded crew. He glanced down and saw me hunch behind the net reel to block the wave that had just breeched our starboard.

"WHAT, what are you doing on deck without raingear on! WHY are you out here without… Do you know how cold this water is! Do you know how quick you can get hypothermia! YOU BETTER GET YOUR GEAR ON UNLESS YOU WANT TO FUCKING DIE!" Charlie screamed.

He wasn't the type of man to raise his voice in anger, especially with me. We had that common bond of growing up south of the Mason-Dixon and would often talk about sweet tea and books by Larry McMurtry. Something was definitely wrong.

I snatched my raingear and scurried back inside. The waves looked to be getting close to twenty feet. Observer protocol states that we should not sample when waves get this high. The pitch and fall of the boat plays hell with your scales even on a calm day. Anything rougher than fifteen feet and reading the scales quickly becomes a futile task. Besides, it's just plain dangerous.

I went up to the wheelhouse to tell Charlie that I wasn't going to sample this haul. More important, I wanted to look him in the eye and make sure we were square. I've never liked anyone yelling at me, even if he was right.

"The cables are crossed; we can't get the net up. We're sitting dead in the water, and the weather's starting to get out of hand. I got one guy out there that's too old to stand and one that's too young to know anything," Charlie said somewhat dejectedly. "I need you up here with me. I can't watch those guys and steer us through this shit at the same time. Just keep an eye on the boys and make sure no one gets washed over." He paused and looked at me. "This is a bad situation, Ben."

I didn't doubt him. The wheelhouse is located in the very front of the boat and sits higher than the rest of the ship. It is enclosed with glass, providing a panoramic view of the sea. The sun hadn't quite set, and I could see what we were up against. I had never seen the ocean behave so violently. Waves crashed into one another from all sides just as the water does inside a washing machine. The hundreds of birds that constantly followed the boat in search of an easy meal had all disappeared. The boat pitched and plunged in all directions. Then it got rough.

The winds had reached gale force and were still gaining strength. Forty-foot swells tossed the *Caravelle* up into the Alaskan night at a steady pace, and gravity is not lost on the sea. If you go forty feet up, you've got to go forty feet back down in order to abide by the laws of the universe.

I watched helplessly as the crew struggled with the rigging until finally the motion of the sea freed the lines from each other. The fisherman hastily began to pull the net back in. Charlie muttered something, more to himself than to me, about how the boat wasn't designed to take this abuse, how it was made to sail in the calmer waters of the Gulf of Mexico. It was exactly the last thing I needed to hear.

I saw it coming, but my mind somehow couldn't quite grasp what my eyes were seeing in those brief moments. A breaker, fifty feet in height, was making its way right for our stern. Breakers, or rogue waves, differ from swells in that they collapse on themselves like waves you see at the beach. I wanted to say something, wanted to warn the crew, but the wave came out of nowhere and came fast.

James saw it first, and as if in some desperately choreographed procedure he and Gordon made a mad dash toward the wheelhouse. They leaped into the air at seemingly the same moment and clung to the two ladders that lead to the upper deck.

John froze, his mind trying to process what was transpiring. He'd seen James and Gordon race past him and make for the wheelhouse. So, as he had done his whole life, he headed in the same direction as his brother, but he was too late. The giant wave broke over the stern and caught him while he was in the middle of the deck. The water traveled with such force that it smacked against the window I was looking through, which was a good fifty feet from the back of the boat where it had first hit. The deck itself was covered with three feet of water for several seconds. James and Gordon had made it to higher ground, but poor John was not in sight. When the water receded, he popped up from under the net reel, coughing and spitting up the cold sea, his face a snapshot of fear and his life still whirling in front of his eyes. Both his will and his desire for a paycheck had been washed overboard.

Charlie screamed from the wheelhouse in an effort to rally his crew. It wasn't about money anymore. They had the first section of the net secured on board, but twenty yards and twenty metric tons of fish were still out in the churning madness trying to pull the boat from Dixie down to Davy Jones's locker.

John just stood there listlessly as James and Gordon tried to empty the codend. Charlie screamed at the crew again, this time directing his words at John, trying to light a fire in him that might burn away the kid's fear. But John couldn't keep his eyes from the surrounding sea. He knew he hadn't developed the sixth sense that his brother and Gordon had that would warn them of impending danger.

"He's lost it. Poor kid's worthless now," Charlie said to me.

I looked out at the crew. The boat pitched and plunged so dramatically that one second I was looking up at them, the next I was looking down. Gordon slipped, fell, and hopped back up.

"I don't mind going out and shoveling some fish if you need me to," I said to Charlie. He took a step closer to me and looked me dead in the eye. A skipper, by honor and by law, is responsible for every person on his ship. Observers are government property, on board only to collect biological data, and he knew he would have a big pile of shit to clean off his shoes if something happened to me.

"For the integrity of the boat and our lives, I'm going to let you," he said. "I want you to run the hose. Let them do the grunt work; that's what they're getting paid to do." I jumped down to the galley to gear up.

The inside of the boat was a small disaster area. Coffee cups, pencils, cigarette lighters, and magazines slid around on the floor. Our clothes and sleeping bags had been slung from our bunks and piled up in the hallway. As I put on my raingear, a jar of homemade jelly that Charlie's wife had made fell out of the cabinet and broke on the floor.

I suited up tight, careful to button every button on my jacket. I duct-taped the cuffs of my bright orange waterproof bibs to my XtraTuffs so water couldn't wash up into my boots. I slipped on my gloves, snapped my hood on, and headed out on deck to see what I was made of.

I opened the door, and all my senses were overwhelmed. The sounds of the waves crashing; the smell of saltwater mixed with fish; the feel of a hard, never-ending icy wind; and the silhouettes of the crew dwarfed by the giant swells all sent me into sensory overload.

They looked confused at first, wondering what I was doing out on deck. No observer would ever try to sample in this weather.

"You boys need a hand?" I asked as I choked down my fear. Their confused looks became even more perplexed and then melted into smiles.

"Why, hell yes!" James grinned. In that moment I transcended the role of government employee, sent to spy and meddle in the fishermen's livelihood, and I became a crew member of the *Caravelle*.

"Let Ben run the hose! The rest of you got to bust ass so we can get to some shelter!" Charlie yelled from the wheelhouse.

I had watched the guys use the hose hundreds of times by now. There was nothing much to it. Pressurized seawater is pumped through a three-inch hose to dislodge the fish from the net.

I slung the hose over my shoulder and dragged it toward the codend. I watched as waves thirty feet in height rolled past me so close that I could reach out and touch their blue-green mass. We all struggled to keep our feet under us.

I made my way to the ship's stern. Only a third of the codend was secured on board; the other two-thirds was still out in the chaos, being thrown around at will by the sea. I climbed on top of the codend and straddled it. I've never ridden a Harley or been crazy enough to saddle up on a bull, and so the raw power that was between my legs at that moment was like nothing I had ever experienced, or probably ever will again. I drove my left hand into the net, wrapped my fingers in the mesh, and held on for all I was worth. I draped the hose over my right shoulder and pointed the steady stream of water at the mound of fish.

What stands out now in my memory was the constant motion. The forty foot up, forty foot down of the boat's journey through the storm, the way the codend rolled and twitched back and forth frantically on deck like a giant fish caught in a net.

I heard James yell something I couldn't quite understand. It came from deep within him, a guttural sound that can only be spawned by primal fears. And then another rogue wave broke over our starboard. I buried my face in the net, cheek to cheek with cold rope and colder fish. It was like being blindsided by a frozen 250-pound linebacker. The water was so frigid that my head was instantly struck with brain freeze, like a kid who's eaten ice cream too fast.

I saw Charlie pop out of the wheelhouse and do a quick head count to make sure we were all still on board and hadn't been taken by the sea. And so life went for the next hour. We dodged more rogues, weathered gale-force winds, and stomached swells that were taller than the two-story house I had grown up in somewhere far away from here.

When the last fish was shoveled below deck and into the tanks, we all grinned like giddy schoolboys and headed back inside. As I took off my raingear, I noticed Gordon standing behind me, silently waiting. I took off my XtraTuffs, and without a word he took them from my hand. He headed for the engine room with my boots, and I saw him reach in and pull out the wet, nasty insulated sole inserts.

"I'm gonna take 'em down to the engine room. They'll dry out a lot quicker with all the heat down there. You'll see; they'll be as dry as a bone," Gordon said, smiling.

A warm flush pumped its way through my heart. I had earned their respect.

We dined that night on James's special recipe of ramen noodles, hard-boiled eggs, and Spam. James and I swapped stories of our past; John swore that this was his last fishing trip and decided he'd better just stay home and make sure his girlfriend didn't run off with a dock-worker; Gordon found a nap waiting for him in the wheelhouse.

A few weeks later the fishing season ended and I headed back home.

I've been back in the lower 48 for three years, and I can honestly tell you that not a day has gone by that I haven't thought about the time I spent in Alaska. And I still find it difficult to speak to others about the experiences I had up there. It's not because they were traumatic or because the details have gotten buried in my brain; it's just that I can't seem to describe it all without monopolizing conversations, and being the southerner that I am, I feel like I'm "putting on airs." So instead, I've begun to write about it, now that time has seasoned the stories, allowed me to ruminate and reminisce, and given me a deep longing to relive them.

I recently attended a meeting of the Alabama Ornithological Union that was held out on Dauphin Island. On the way there I passed a sign that read, "Bayou La Batre, 10 Miles." I smiled, thinking about the *Caravelle* and that old cliché about how small the world is. My memories of Alaska, I realized, are never far from my mind, and they remain as palatable and salty as a forty-knot wind.

37

Dependence Day

JEFF BEANE

In late afternoon on the Fourth of July I stood near the summit of the highest point in eastern North America and looked down. There seemed no more fitting a place for me to be on this day—no better vantage point from which to look out across the land where I had spent my few decades of existence. Looking eastward across the state of my birth, I felt that I could see all of it, from the escarpment at my feet to the barrier islands. I could see the salt marshes, maritime forests, pocosins, and river swamps of the coastal plain; my beloved sandhills, pine flatwoods, and Carolina bays; the rolling hills and remnant mixed forests of the Piedmont; and the icy streams, laurel thickets, bogs, and shaded slopes of the fir-covered peaks where I stood. I could see the cancers of Raleigh, Durham, Cary, Greensboro, Winston-Salem, Charlotte, Fayette-Nam, and all the others, and the mindless, death-dealing arteries of the interstate highways. I could see all the creatures I had devoted most of my life to understanding and fighting for: the Green Salamanders in their shaded rock crevices; the tiny Bog Turtles buried in the muck of their remnant sedge meadows; the Gopher Frogs and Tiger Salamanders down in their burrows beneath the pines, awaiting the rains that would drive them in another eight months toward their few remaining breeding ponds; the little southern Hognose Snakes, inching toward unheralded extinction on their dwindling sand ridges; the massive, tenacious Loggerheads, still hauling their bulk ashore to drop their best shot at a future into holes in the sand; and those marvels of efficiency with buzzers on their tails, still hanging on in the face of overwhelming persecution.

But if I could see all these in my mind's eye, there were others I could perceive only as ghosts: the elk and the bison, the cougar and the wolf, the pigeon and the parakeet, the vast stands of mature chestnut, and—even from way up here—the river frog, last spotted in the state twenty-one years earlier. To see a member of my own species who even knew these were missing, I would have to look hard, and harder still to find one who really cared. I thought of earlier naturalists who had passed this way: Mitchell, Catesby, Gray, Bartram, Michaux, and others long forgotten; those seeking knowledge—or something higher—who might have looked down from these same mountains. What would they think and feel if they were here now? Looking out over the sea of red spruce and Fraser fir, I could almost imagine the land as it once was. Almost. The visible scars of roads, power-line cuts, and buildings; the scattered stands of dead and dying firs; and the shouts of human voices echoing from nearby trails would not permit me the view I wanted, and needed.

The waning afternoon air grew cool. In a sunny patch of bluets and St. John's wort along the Commissary Ridge Trail, my friend and I sat on a rock outcrop and soaked up the stored warmth like giant lizards (there being none present to compete with that high up) while we reevaluated the American work ethic. We had come a long way in two hundred twenty years, having hammered our country into what someone surely must have wanted it to be. Still, chaotic life fought its way along in its myriad forms despite our efforts to control or eliminate it, even up here in the clouds. Juncos and waxwings sang, a distant raven croaked, and a hermit thrush floated what Aldo Leopold had once called "silver chords from impenetrable shadows" across the peaks. Grass-of-Parnassus made ready to bloom beside purple fringed orchids. The little Michaux's saxifrage pushed its white flowerheads right out of the bare rock where we sat. And some of the creatures Fate had assigned me as favorites were here: Blackbelly, Spring, and Blue Ridge Two-lined Salamanders inhabited these highest of streams, and everywhere were Carolina Mountain duskies—we found one small log sheltering five clutches of their tiny eggs, each attended by a female. Even two snakes—Ringneck and Garter—made their homes in this cold, high place, and we had seen both that afternoon.

As evening approached, we walked along the trail, speaking little. I saw my companion pick up and pocket a discarded Starburst candy wrapper that lay on the ground.

For most of my life I had known that there was no such thing as independence, and the very concept of it had long amused and confused me. I remembered a day exactly twenty years earlier, when my country was celebrating a bicentennial of what its citizens evidently regarded as freedom. I had spent that Fourth waiting for my great-uncle (one family rumor maintained that he was actually my grandfather) to finish dying so we could bury him. A misfit and misplaced teenager then, I spent long hours pondering and contemplating my life and what, if anything, it meant. On that day I had plenty of time for such idle thoughts, and I can recall thinking how folks ought to instead celebrate a "Dependence Day," an "Interdependence Day," or an "All Things Depend on Other Things Day," not once a year but every single day of their existence.

Two decades and many joys and sorrows later, it could easily be seen that I had not always managed to do that. Many days had not seemed worth celebrating, and I had plodded along, trapped in artificial and sometimes miserable environments and situations. The ability to love life is surely a blessing, but I sometimes believe that loving it too much—at least in this place and time—is nearer to a curse. Cherishing life regardless of what form it happens to take always seemed to me the logical road to peace and happiness. Why, then, could I come close to finding those things only in a shrinking number of places? If all life was sacred, and humans really were as much a part of nature as anything else, why could I not rejoice in their abundance and activities? Why was it that the miracle of a manicured lawn, fescue plot, toy poodle, cellular phone, sports arena, or shopping center brought me no sense of joy and wonder? Why was it that I needed places with plants whose names I did not know and animals that could not be bred in confinement? And why did I need the reassurance that others required the same places for the same reasons? Did they, in fact, need them as much as I and not realize it? And could I ever play even a small part in helping those who had lost touch with their true needs to regain the values necessary to sustain us all? Up on Mount Mitchell that day, I thought I might have touched something that felt like an answer, but if so, it slipped out of my hands as easily as a *Desmognathus* might, leaving me to recall the hard words of the eccentric artist Walter Ingles Anderson: "Those who have identified with Nature must pay the consequences."

That night was the coldest I had ever known in July. From the summit after dark we saw in all directions the distant fireworks displays of a dozen towns celebrating an illusion, touting the greatness of a country they knew little or nothing about. The call of just one saw-whet owl would have better fit the kind of celebration we craved.

The next day we would descend, to look for things—things we needed. Under the guise of studying them, maybe, but really just because we needed to see that they still existed how and where they were supposed to: the Bog Turtles, the Green Salamanders, and the other specialized creatures whose tour of duty in the great war for survival was almost over but whose continued existence fueled our own, and from whose well-being we were anything but independent.

38

The Old Camp Coffee-Pot

BADGER CLARK

Old camp-mate, black and rough to see
A hard-worked aid and ally you
In all of my single-handed wars
With naked nature's savagery
Your scars are marks of service true,
Dear loving-cup of out-o'-doors,
And history in every spot
Has battered you, old coffee-pot.

Oh, black Pandora-box of dreams!
Though dry of drink for mortal needs,
Out of your spout what fancies flow!
The flash of trout in sunny streams,
The swoop of ducks among the reeds,
The buck that paws the reddened snow—
What suns and storms, what dust and mire,
What gay, tanned faces round the fire!

So, vividly as clouds that blaze
Above a sunset's rainy red,
Scene after scene, you bring to me
The camps and trails of other days.
And as a shell, long dry and dead,
Holds echoes of its native sea,
So dear old murmurs, half forgot,
Rise from your depths, old coffee-pot.

I hear the stir of horse's hoofs,
The solemn challenge of the owl,
The wind song on the piny height,
The lilt of rain on canvas roofs,
The far-off coyote's hungry howl,
And all of the camp sounds of the night.
They rise—a thousand things like these—
From you, old well of memories.

Our fires are dead on hill and plain
And old camp faces lost and gone,
But yet we two are left, old friend.
And as the summers bloom and wane
May I meet you at dusk and dawn
By many fires before the end,
And drink to you in nectar hot
From your black throat, old coffee-pot.

Acknowledgments

I'd like to express my appreciation to hundreds of people around the globe who responded to the idea for this anthology with enthusiasm and encouragement, especially those who went out on a limb to send me stories. I hope you all will continue to record your experiences and pass them on, as these tales preserve a heritage of field biology that mere data could never reveal.

I am forever grateful to Jeff Grathwohl, director and senior editor of the University of Utah Press, because he believed in this book from the start and polished it up so that it would shine.

I've got to thank my husband, Chris, too—even though he knows. He is my partner in the field, in life, and in a wild new venture called parenthood—we are one hell of a team. *The Back Road to Crazy* is just one of many dreams we've hashed out over Saturday breakfasts through the years, and he gave me everything I needed to see it through.

Finally, I reserve a most special acknowledgment for our beautiful daughters, Rita and Sophia. In them, I have discovered a whole world of joy, adventure, and purpose I never imagined.

Contributors

Jeff Beane was born in Asheboro, North Carolina, and has had a life-long interest in all aspects of natural history and conservation, especially the natural history, zoogeography, and conservation of reptiles and amphibians in the Southeast and longleaf pine ecosystem ecology. He has been employed by the North Carolina State Museum of Natural Sciences in Raleigh since 1985, and his writings have appeared in a variety of popular and scientific publications. He took the back road (some might say the main highway) to crazy long ago.

Jennifer Bové obtained her B.S. in biology from the University of Missouri in 1996 and has traveled cross-country to pursue fisheries and wildlife work. A native of the Midwest, she now lives in Washington state. Currently, Jennifer spends most of her time raising two daughters, writing, and developing an ecology-based board-game company called Green Apple Games. Her writing has appeared in the *Missouri Conservationist*, *Wild Outdoor World*, *Women in Natural Resources*, and *Heart Shots*, a hunting anthology edited by Mary Zeiss-Stange.

Alice Cascorbi studied biology at Carleton College, science writing at U.C. Santa Cruz, and conservation biology at the University of Minnesota. Thanks to various lab and field positions, she knows how to sequence oat DNA, find songbird nests, mark and recapture goldenrod leaf beetles, and burn prairie. She is currently a researcher-writer for the Monterey Bay Aquarium's Seafood Watch program, evaluating the ecological sustainability of commercial fisheries.

Badger Clark (1883-1957) was named the first poet laureate of South Dakota and is remembered for his ballads of the American West. According to the Badger Clark Memorial Society, Clark devoted much of his life to studying, cultivating many close friendships, and most important, enjoying the great out-of-doors.

Troy Davis has worked as a biologist for the U.S. Fish and Wildlife Service and the Bureau of Land Management and is currently a ranger for Yellowstone National Park. His wildlife experiences all over the United States have inspired him to search for a graduate studies program that will keep him in the mountains.

Elizabeth Dayton is a writer and educator who has worked as a field biologist in the Pacific Northwest, southern California, North Carolina, west Texas, and Mexico. She received her B.S. in zoology from Humboldt State University in Arcata, California, and her M.A. in education from Texas A&M University. She currently lives in Texas with her husband and two daughters.

Barbara Blanchard DeWolfe, a third-generation Californian, was born in San Francisco. After earning her Ph.D. from the Museum of Vertebrate Zoology at U.C. Berkeley, she taught zoology for thirty-one years at U.C. Santa Barbara, retiring in 1977.

Eric Dinerstein is chief scientist and vice president for science at World Wildlife Fund-U.S. He is the author of a recent book on the successful recovery of the greater one-horned rhinoceros, an endangered species of Asia, entitled *The Return of the Unicorns,* published by Columbia University Press. The chapter in this volume is excerpted from the forthcoming *Tigerland, and Other Unintended Destinations,* to be published in 2005 by Shearwater Press.

Joseph L. Ebersole, a lifelong lover of watery secrets and hidden possibility, is a stream fish ecologist currently working for the U.S. Environmental Protection Agency. He lives with his wife, Elizabeth, and sons Levi and Alexi in Corvallis, Oregon.

Wendell R. Haag is a lifelong avid naturalist and has studied the ecology of freshwater mussels and fishes for nearly twenty years. A native of Kentucky, he is currently employed as a research fishery biologist with the Forest Service's Center for Bottomland Hardwoods Research, Oxford, Mississippi.

Tonya M. Haff began a career in field ornithology in 1995 and has focused on riparian bird monitoring and conservation. She received an M.S. in resource ecology from the University of Michigan in 2001 and is now curator of a natural history collection in Santa Cruz, California. Her current interests are ornithology, human foraging ecology, and mycology.

Sharyn Hedrick went from studying the largest sea creatures to the smallest: phytoplankton. The artist in her couldn't resist the intricate designs of these microscopic plants. She has been with the Smithsonian Environmental Research Center for the last sixteen years studying diatoms, dinoflagellates, and the like and the effects of light, nutrients, and turbidity on their growth.

Betsy L. Howell is a writer and wildlife biologist who lives on Washington's Olympic Peninsula. She is currently working on a memoir about how war has affected her life through the battle experiences of her father and great-great-grandfather.

James Lazell founded the Conservation Agency, a scientific research nonprofit, in 1980. He has published more than two hundred peer-reviewed papers, many popular articles, and three books.

Ran Levy-Yamamori is an ecologist as well as a natural history writer and illustrator. He develops educational tools such as flower identification keys and environmental education programs. He wrote *Wild Flowers of Japan* and coauthored *Garden Plants of Japan* and *Flowers of the Eastern Mediterranean*. He also operates Har-Yaar Books, a small publishing house for children's natural history books.

David M. Liberty is a father, husband, son, and grandson. He lives in Hood River, Oregon, and works for the StreamNet Library in Portland as the assistant librarian. He is also a musician, artist, craftsman, and mechanic who writes once in a while. He currently co-hosts Native Nations on Multnomah Community TV and Mitakuye Oyasin (All My Relations) on KBOO FM. David is an enrolled member of the Confederated Tribes of the Umatilla Indian Reservation.

Patrick Loafman has been a seasonal wildlife biologist since graduating from Auburn University in 1988. He divides his time between wildlife research in the summer and writing in the winter. He has two published chapbooks of poetry, an unpublished manuscript of nature essays, two almost-finished novels, and box loads of short stories.

Mark W. Moffett is a high-school dropout who went on to complete a Ph.D. under Edward O. Wilson at Harvard. He studies animal behavior and the physical structure of ecosystems such as forest canopies and is a frequent explorer and photojournalist for *National Geographic* magazine.

Dan Mulhern received his M.S. degree from Kansas State University in 1984 and began working for the Fish and Wildlife Service in Manhattan, Kansas, that same year. Since 1988 he has been the endangered species coordinator for the Service's activities in Kansas. His wife, Terry, is an artist, and they have three children.

Ram Papish (www.rampapish.com) splits his time between fieldwork and artwork. He is known for his habits of painting wildlife on blue jeans, drawing on envelopes, and giving lively and entertaining karaoke performances. He was the 2004 International Migratory Bird Day artist.

Michael Rogner is a biologist for the Point Reyes Bird Observatory and is currently studying avian response to riparian revegetation along the Sacramento River. His work has appeared in a number of publications, including *The Quarterly*, *Northwest Review*, and *Elysian Fields Quarterly*.

Lynn Sainsbury has worked as a seasonal biologist/forester with the Forest Service for the last fifteen years. Her winters are spent in a small house overlooking the Potomac Valley in western Montana. One newly adopted son and two long-term companion cats keep her both busy and in good company.

Charles F. Saylor was born in east Tennessee and graduated from the University of Tennessee, Knoxville, in 1972 with a B.S. degree in zoology. He has spent the last thirty-two years working on various aquatic biology projects for the Tennessee Valley Authority.

John Shivik is a research wildlife biologist who is still trying to find better ways for predators and humans to coexist. He continues to bust new dance moves with coyotes and an occasional wolf or bear. He lives in Logan, Utah.

Chris Smith worked more than twenty-two years for the Alaska Department of Fish and Game, where he survived numerous adventures besides those reported here. He is now chief of staff for Montana Fish, Wildlife, and Parks, and both his life and his job seem tame by comparison.

Scott Stollery earned an M.A. in English literature at Northern Arizona University in 1998. After a brief stint teaching English overseas, Scott decided he was better suited to working outdoors and jumped ship to environmental fieldwork. This work, predominantly with plants and birds, has taken him across the United States and has provided him with a steady stream of inspiration for his writing, music, and art.

Benton Taylor is a native of Harvest, Alabama. After completing two contracts as an NMFS groundfish observer, he was employed as a field technician with Avifauna Northwest and the Alabama Black Bear Alliance. He is currently working as a GIS analyst and hopes to return to the field soon.

Cameron Walker graduated from U.C. Berkeley with a degree in bioresource sciences in 1999 and received a graduate degree in science writing from U.C. Santa Cruz. Since then she has lived around the West and held a number of odd jobs. She has written for *Discover*, *Outside*, and *California Wild*.

Howard Whiteman is a native of Pittsburgh, an ardent fan of the Steelers and heavy metal, a hunter and fisher, and a graduate of Allegheny College (B.S.) and Purdue University (Ph.D.). Currently, he's an associate professor of biological sciences at Murray State University, Murray, Kentucky, where he teaches ecology and studies the evolutionary ecology and conservation biology of amphibians. Howard lives in a log home with his fellow biologist wife, two wonderful children, and a stray dog.

Roy A. Woodward was born and raised in Fort Bragg in the coast redwood region of California. He received a B.S. and M.S. from Brigham Young University and a Ph.D. in botany from U.C. Davis. He has worked for several federal and state agencies, spent seven years with Bechtel Corporation working on environmental projects on six continents, and currently is head of the vegetation and wildlife monitoring program for California State Parks. His wife and three children often accompany him during his fieldwork, and he enjoys fly-fishing, serving as a Boy Scout leader, and playing the piano.